AF413065

Microfabricated Sensors

ACS SYMPOSIUM SERIES **815**

Microfabricated Sensors

Application of Optical Technology for DNA Analysis

Richard Kordal, Editor
University of Cincinnati

Arthur Usmani, Editor
Altec USA

Wai Tak Law, Editor
PortaScience Inc.

American Chemical Society, Washington, DC

Library of Congress Cataloging-in-Publication Data

Microfabricated sensors : application of optical technology for DNA analysis / Richard Kordal, Arthur Usmani, Wai Tak Law, editors.

p. cm.—(ACS symposium series ; 815)

Includes bibliographical references and index.

ISBN 0–8412–3763–8

1. DNA microarrays—Congresses. 2. Biochips— Congresses. 3 Biosensors—Congresses.

I. Kordal, Richard Joseph. II. Usmani, Arthur M., 1940- III. Law, Wai Tak, 1951- IV. American Chemical Society. Division of Industrial and Engineering Chemistry. V. American Chemical Society. Meeting. (219th : 2000 : San Francisco, Calif.) VI. Series.

QP624.5.D726 M53   2002
610′.28—dc21   2001056737

The paper used in this publication meets the minimum requirements of American National Standard for Information Sciences—Permanence of Paper for Printed Library Materials, ANSI Z39.48–1984.

PRINTED IN THE UNITED STATES OF AMERICA

Foreword

The ACS Symposium Series was first published in 1974 to provide a mechanism for publishing symposia quickly in book form. The purpose of the series is to publish timely, comprehensive books developed from ACS sponsored symposia based on current scientific research. Occasion-ally, books are developed from symposia sponsored by other organizations when the topic is of keen interest to the chemistry audience.

Before agreeing to publish a book, the proposed table of contents is reviewed for appropriate and comprehensive coverage and for interest to the audience. Some papers may be excluded to better focus the book; others may be added to provide comprehensiveness. When appropriate, overview or introductory chapters are added. Drafts of chapters are peer-reviewed prior to final acceptance or rejection, and manuscripts are prepared in camera-ready format.

As a rule, only original research papers and original review papers are included in the volumes. Verbatim reproductions of previously published papers are not accepted.

ACS Books Department

Contents

Indexes

Preface

Since their discovery DNA-or BioChips have been adopted by the research community at a stunning rate. This is just one demonstration of the importance of this new technology to the scientific community. These chips are enabling scientists to conduct experiments they only dreamed of performing previously. For example, in one application these chips enable researchers to monitor the expression of literally tens of thousands of genes simultaneously and to determine how complete batteries of genes coordinate their activity. The scientific implications of this are enormous. In addition, the chips are the basis of a completely new field of medicine, namely pharmacogomics. DNA- or BioChips are finding ever increasing use in medical diagnostics, drug discovery, and bio/analytical chemistry.

This volume has been organized in three sections; namely, DNA chips, detection technologies, and optical sensing and processing. This book should be useful to life scientists in the biomedical field. Bioengineers or nanotechnologists should also find this volume worthwhile.

Richard Kordal
University of Cincinnati
3223 Eden Avenue
G–07 Wherry Hall
Cincinnati, OH 45267

Wai Tak Law
PortaScience Inc.
337 Tom Brown Road
Moorestown, NJ 08057

Arthur Usmani
Altec USA
4101 East 30th Street
Indianapolis, IN 46204

DNA Chips

Chapter 1

Gene Expression Microarrays: A Platform for Discovery in Biological Systems

Karen Seta, Hie-Won Kim, Yong Yuan, Gang Lu, Zachary Spicer, Richard Kim, Tsuneo Ferguson, Peterson Pathrose, and David E. Millhorn*

Genome Research Institute and Department of Molecular Physiology, 231 Albert Sabin Way, University of Cincinnati, Cincinnati, OH 45267–0576

The Human Genome Project and various other genome sequencing projects have led to a need for high throughput approaches to measure and analyze gene expression patterns. Currently, the most broadly used approach for measuring gene expression profiles is high-density spotting of oligos or PCR products onto glass slides. These slides are then used as a platform for competitive hybridization of fluorescently-labelled cDNAs from control and experimental cells or tissue. Here we shall describe our system for performing gene expression profile analysis and examples of how this system has been used to better understand complex physiological mechanisms and disease processes.

INTRODUCTION

In recent years considerable progress has been made in the diagnosis and treatment of many diseases, and in the understanding of complex physiological mechanisms in healthy cells. Much of this progress has resulted from years of research on the genetic and environmental factors that regulate cell function. The on-going quest to understand the role of gene expression in mediating alterations in cell phenotype has historically been restricted to the study of single genes and proteins. Yet, most complex biological functions and disease processes (such as cancer) involve the simultaneous expression of hundreds or even thousands of genes and proteins (Figure 1). Recent technological advances have made it possible to analyze global alterations in mRNA expression. Included among these advances is the establishment of large sequence databases, which have resulted from the various genome-sequencing projects, and the development of high-density nucleic acid microarrays. Investigators are now empowered with the ability to identify transcription (mRNA) profiles that correspond to changes in cell function and phenotype. This line of investigation is referred to as "genomics."

Functional genomics consists of experimental approaches to identify genes that mediate both normal cellular functions and disease processes. Gene chip (cDNA microarray) technology and the expanding gene sequence databases have made it possible to identify the complement of transcribed genes (the "transcriptome") within cells and tissues that are expressed during development or in response to specific stimuli (1-6). The gene expression profile that results from extra-cellular stimuli and genetic mutations are analyzed by various computational approaches and then classified according to our expanding knowledge of gene structure and function. Thus, functional genomics offers the potential to markedly expand our understanding of biological systems and how these systems respond to environmental stimuli. Genomic approaches promise to accelerate our progress in deciphering the molecular basis of complex physiological systems and genetically acquired diseases. Observations made with these approaches will allow investigators to generate sophisticated hypotheses, which will greatly enhance their research. Here, we shall describe our approach for measuring gene expression profiles in living cells, which features current state-of-the-art technologies and methods for microarray analysis.

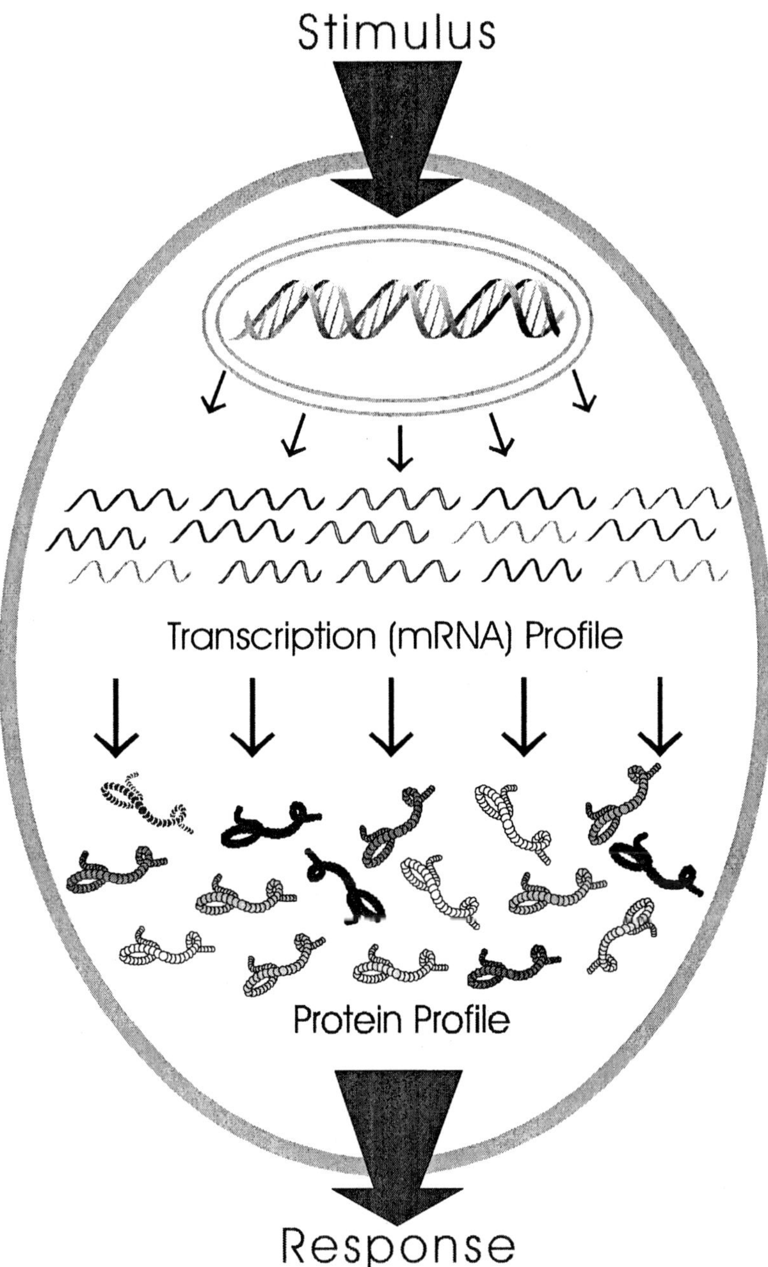

Figure1. *Genomics and proteomics response to an extracellular stimulus.*

METHODS

Selection of cDNAs for Producing Microarrays

After considerable "proof-of-concept" testing, we made the decision to use custom subtracted cDNA libraries for making microarray templates (cDNAs or PCR products micro-spotted onto either glass or nylon supports) for gene expression profile analysis. Traditional cDNA libraries are not the best choice because of the high redundancy of clones, and because of the high titer (500,000 to 1,500,000 clones). It is far too expensive and time consuming to screen cDNA libraries for suitable clones (e.g. non-redundant) to use for microarray analysis. The subtracted cDNA libraries we make use a modified subtractive suppression hybridization (SSH) technique (Clontech Laboratories, Palo Alto, CA) (7-9). We have used this approach to make more than 30 very high quality SSH libraries that are used for making templates for microarray analysis. Quality control experiments performed on each library revealed very little redundancy (<5%), and importantly, sequencing revealed the presence of clones of interest. The titer (4,000 - 30,000 clones in 1 ml of bacterial stock) of the subtracted libraries is ideal for microarray analysis. Perhaps the most important feature of this particular approach is that it provides the investigator with a subset of the genome that contains his/her genes of interest, both novel and known. Another practical advantage of subtractive libraries is that they can be made from any species, which obviates the concern of using the human and mouse commercial clone sets for other species (e.g. rat).

Subtracted cDNA libraries provide a powerful biological approach for obtaining clones that are differentially expressed between two populations. Two recently developed PCR-based methods, subtractive supression hybridization (SSH) and representational differences analysis (RDA), have been used successfully to clone genes which are differentially expressed between control and test cells or tissue (10-13). We have had very good success in making subtractive cDNA libraries using the SSH approach, which combines a high subtraction efficiency with an equalized representation of differentially expressed mRNAs (7, 12). SSH is based on a specific form of PCR that permits exponential amplification of cDNAs which differ in abundance, whereas amplification of mRNAs of identical abundance in the control and experimental populations is suppressed (7) (CLONTECH PCR-Select, Palo Alto, CA). Figure 2 illustrates the major steps in the production of SSH subtracted libraries.

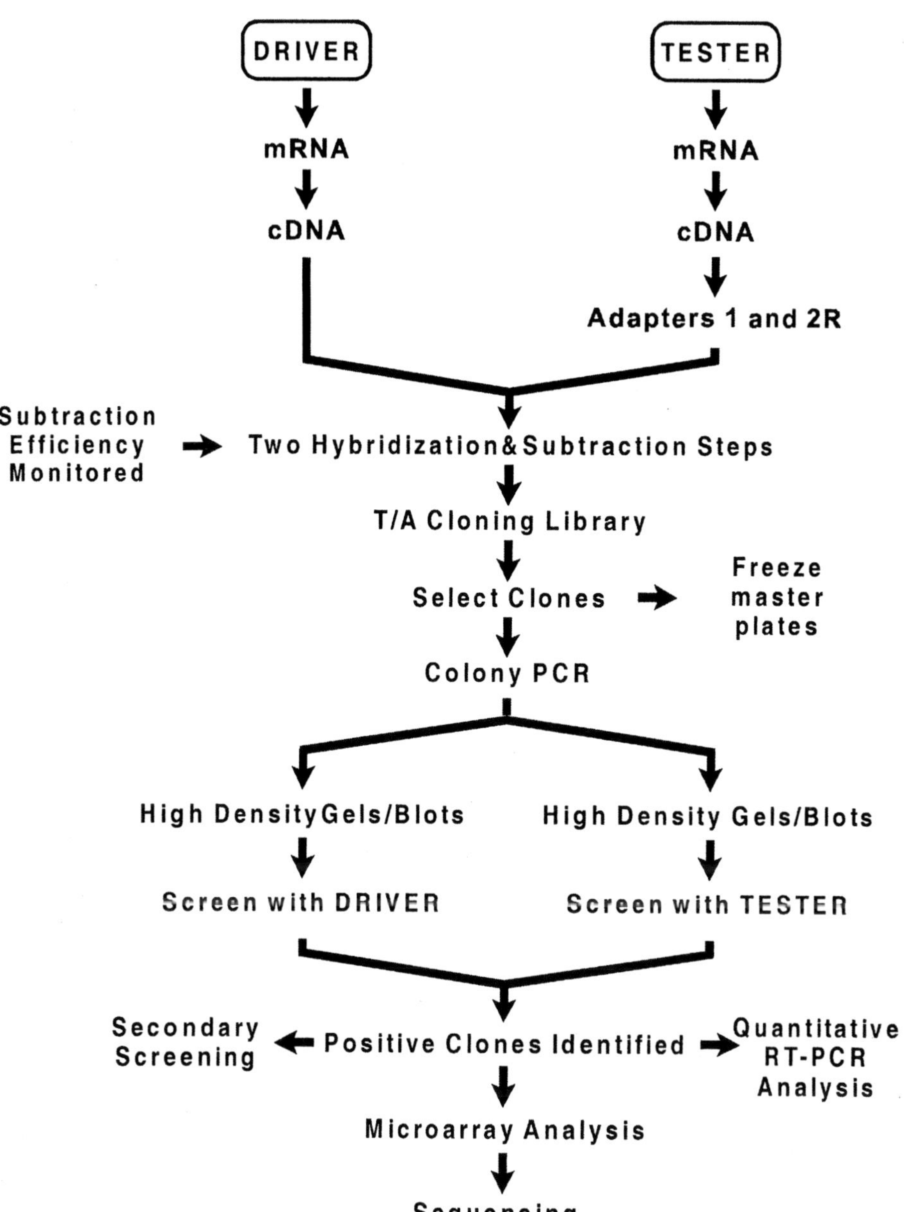

Figure 2*. Steps for making SSH cDNA libraries.*

Another option for obtaining cDNAs for microarray analysis is to purchase clones from a commercial source. These acquired clones are usually sequenced verified and can be "sub-arrayed" to create user-specific microarrays (e.g. a cancer-enriched chip). A major advantage of the subtracted libraries over commercial clone sets is the opportunity to identify novel genes. Moreover, it is entirely possible that the genes that mediate the phenotype of interest are not included in the commercial clone sets, which represent a relatively small portion of the entire genome and are composed primarily of ESTs.

Production and Quality Control of SSH cDNA Libraries

Poly A+ RNA for control ("driver") and experimental ("tester") are made and then quantitated on a formaldehyde gel. Each sample is concentrated to a range of 0.5 to 2 ug/ml. cDNAs are prepared from the two poly A+ RNA samples by reverse transcription. Second strand cDNA synthesis is then performed and the ds cDNAs are digested with a four-base cutting enzyme (Rsa I) that yields blunt ends. The cut ds cDNAs are analyzed on a 1% agarose gel. Following this the tester (i.e. experimental) ds cDNA pool is divided into two equal portions, and the ds cDNAs in one portion are ligated with adaptor 1 and those in the other pool with adaptor 2R using T4 DNA ligase. Since the ends of the adaptors do not have a phosphate group, only one copy of each adaptor attaches to the 5' ends of the cDNA. Following successful ligation of the adaptors, hybridization is performed with excess "driver" added to each "tester" sample. The samples are heat denatured and allowed to anneal. The concentration of high- and low-abundance cDNAs in the tester population is equalized due to second-order hybridization kinetics. Differentially expressed cDNAs in the tester population (which remain single stranded under the conditions used) are also enriched, because cDNAs that are not differentially expressed form hybrids with driver cDNA. A second hybridization is performed by mixing the two primary hybridization samples (without denaturing) with an excess of denatured driver. The tester cDNAs that remained single-stranded in the first hybridization now form tester-tester hybrids that contain both adaptors. This population of hybrid molecules that contains both adaptors (N1 and N2R) is the population that represents the differentially expressed tester sequences. The entire population of hybridized molecules is then subjected to PCR to amplify these desired sequences. In this PCR reaction, driver sequence does not get amplified because it has no adaptors. Tester/driver hybrids are only linearly amplified because they contain adaptor at only one end. Tester/tester hybrids that have the same adaptor at both ends will form hairpin loops under the conditions used, and will not be amplified ("suppression PCR").

The subtracted library is ligated into T/A cloning vector (Invitrogen, Inc.) and electroporated into phage-resistant bacterial cells (DH10B), which are then stored in glycerol at −80° C. An aliquot (100-200 μl) of the library is plated on a LB agar plate with the appropriate antibody for the purpose of determining the titer of the library. The T/A cloning vector has a β-galactosidase site which provides a mechanism for color (blue-white) selection of bacterial colonies that contain a subtractive clone. Positive colonies are inoculated in 96-well plates with antibotic and 10% glycerol and stored at -80C. This becomes the "original" copy of the library. This library is copied once to make a "master" copy and once to make a "working" copy of the library. All libraries are stored in ultra-cold freezers.

A variety of quality control tests are performed on each subtracted cDNA library. First, the cDNA inserts (produced by PCR) from 60 to 80 randomly selected clones are examined on 2% agarose gels. It can be seen that virtually all the clones contain unique inserts and that the size of the inserts range from about 0.3 to 1 kb. Sequencing of the clones revealed low redundancy of clones (<5%). To construct the library, the original cDNA pools are digested with *Rsa*I, a 4-base cutting enzyme, thus 0.3 to 1 kb corresponds to the expected range of inserts.

We routinely perform "virtual" Northern blots to measure expression of various cDNAs in the subtracted library. Virtual Northern blots are made with cDNA generated by reverse transcription from RNA. The cDNA samples are separated on standard DNA agarose gels instead of total or polyA$^+$ RNA on formaldehyde gels, and yield information similar to that provided by standard Northern blots (8).

Importantly, the subtraction technique *excludes* abundant sequences that are not differentially expressed between the "tester" and "driver" mRNA populations. Thus, highly expressed genes that are not regulated (i.e., housekeeping genes), such as glyceraldehyde-3-phosphate dehydrogenase (G3PDH) should be excluded from the subtracted pool. An example of this is shown in Figure 3, which is from a library that we recently made to identify mRNAs that are regulated by hypoxia in PC12 cells. In this example G3PDH levels were analyzed in cDNA samples from the final (subtracted) cDNA library and in the original (unsubtracted) cDNA pool by RT-PCR. Samples were subjected to 18, 23, 28, or 33 cycles of PCR, as indicated in the Figure. It can be seen that G3PDH is readily detectable in the original (unsubtracted) cDNA sample, but is completely absent in the subtracted sample (even after 33 PCR cycles). G3PDH is a highly abundant gene that is not normally regulated by extracellular stimuli (14). Thus, G3PDH, which is not regulated by hypoxia in PC12 cells, was accurately excluded from the library.

To further determine the overall accuracy of the subtraction (*i.e.*, the rate of false positives) a subset of the isolated clones was subjected to a

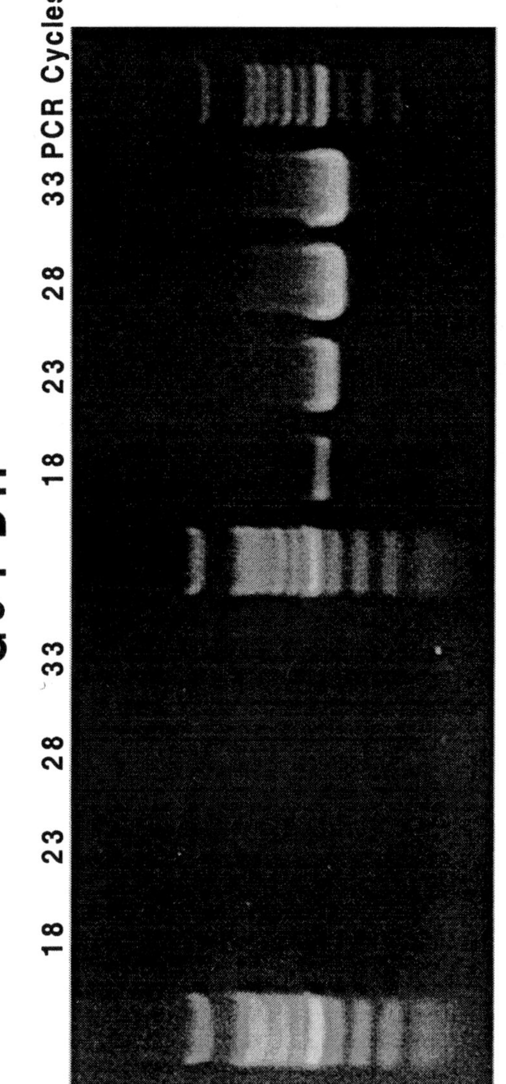

Figure 3. PCR for G3PDH in tester (subtracted) and driver (unsubtracted) samples following two hybridizations. M = marker.

secondary screen by dot blot analysis. Eighty-six clones were randomly selected from the "forward subtracted" (or "up-regulated") library and the inserts were amplified by PCR in a 96-well plate. The resulting PCR products were spotted and fixed on 2 duplicate nylon filters. These filters were then hybridized with ^{32}P-dCTP-labeled cDNA probes made from the "forward subtracted" pool or the "reverse subtracted" pool of cDNA. Arrows in Figure 4 indicate spots that are likely to be "true positives", or cDNAs which were present at higher levels in the "forward subtracted" (hypoxia-induced) pool versus the "reverse subtracted" (hypoxia repressed) pool. In many of the remaining clones the hybridization signal intensity was too low to accurately determine whether these cDNA sequences were significantly enriched in one cDNA pool compared to the other. Nevertheless, it is clear from this experiment that the forward-subtracted library is enriched in sequences that are more abundant in the hypoxia-exposed RNA sample *versus* the control sample. Thus, the subtracted cDNA library approach has enabled us to isolate bona fide hypoxia-regulated genes.

The cDNAs identified by SSH are sequenced and subsequently used to construct custom cDNA microarrays. Each new subtracted library is characterized in this manner. Once this has been performed, the library is stored in glycerol at −80° C until it is used for making probes for printing onto glass slides, which are then used for microarray analysis. It is our practice to sequence and annotate each clone in the library. Alternatively, the investigator can choose to microarray the library and then sequence those clones that are regulated. This approach allows for the economical production of microarrays with a low redundancy of representation, and a relevant set of genes to survey. In addition, this an effective approach for identifying rare and novel genes.

Microarray Production and Hybridization Procedures

The cDNAs of interest are PCR amplified from clones in our subtracted libraries or from acquired clone sets. Following purification and quality control, aliquots (approximately 5 ng of DNA/spot) are printed onto aminosilane coated glass slides using a computer-controlled microarrayer robot (OmniGrid, Gene Machines). The microarrayer is located in a temperature and humidity controlled room in the Genome Research Institute. The DNA samples that are used for hybridization analysis with the microarray printed cDNA template are generated from total RNA populations (control and experimental) and labelled with either green (Cy3) or red (Cy5) using a two-step amino-allyl method (Fair Play Microarray Labelling kit, Stratagene, LaJolla, CA.). Total RNA rather than mRNA is used to maximize the amount of mRNA that can be

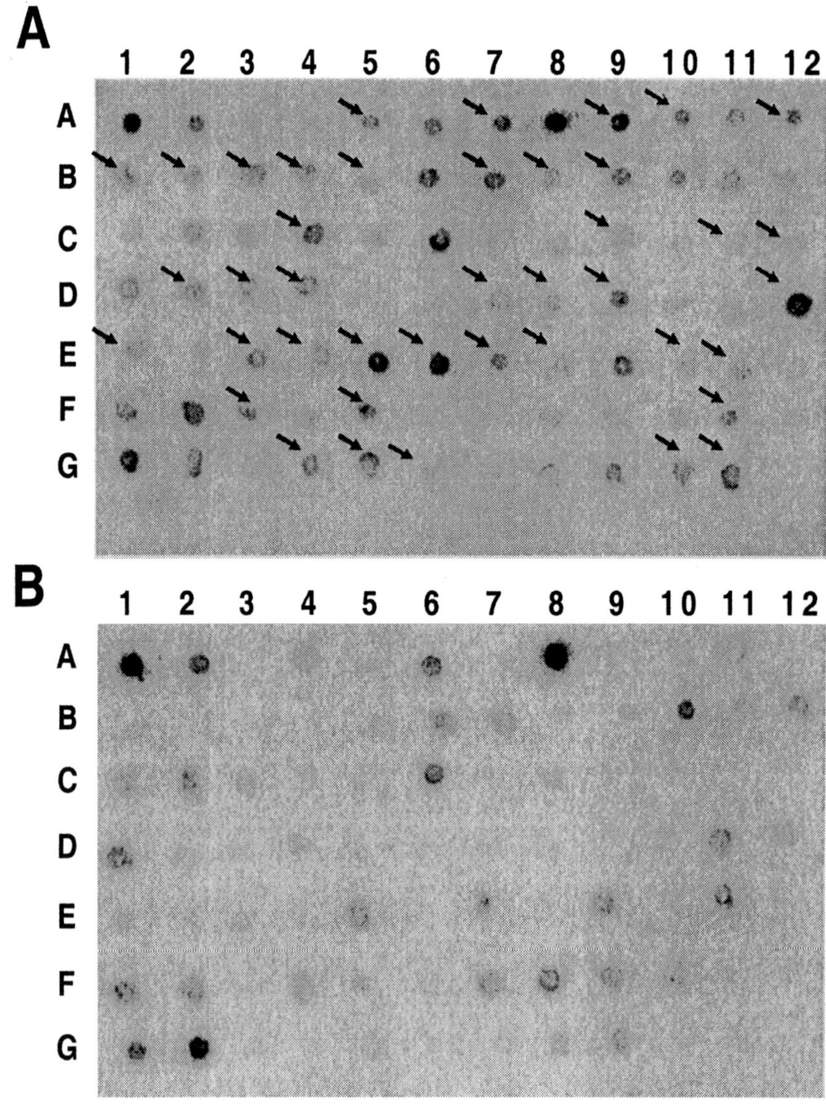

Figure 4. *Dot blot results showing clones that are differentially-regulated by hypoxia in the hypoxia-subtracted cDNA library.*

obtained from a given tissue or cell sample. The fluorochromes (Cy3 and Cy5) have widely separated excitation and emission spectra. The absolute fluorescence intensity for red or green is proportional to the level of expression, with equivalent expression levels yielding a yellow signal. Further, the two-color approach provides consistent across-the-array competitive hybridization controls. The combination of these effects leads to an improved signal-to-noise ratio, which allows the detection of 2-fold differences between mRNAs expressed at low levels. We have used this approach to identify approximately 200 hypoxia responsive cDNAs in an endocrine tumor cell line (pheochromocytoma cells) (12,15).

Steps for Performing a Microarray Experiment

i) Sample Preparation. Clone inserts from the subtracted libraries are amplified by bacterial PCR. The PCR products are purified with the MultiScreen PCR 96-well Filtration System (Millipore, Bedford, MA). DNA concentration is measured in a μQuant 96-well format UV spectrophotometer (Bio-Tek Instruments). DNA yield is calculated and the samples are transferred to V-bottom microtiter plates for printing. The samples are dried in a Speed-Vac and resuspended in 3X SSC (spotting solution) so that the final concentration is 0.1 - 1 mg/ml. PCR products are spotted onto aminosilane-coated slides (TeleChem, Sunnyvale, CA) using an OmniGrid™ robot (GeneMachines, San Carlos, CA) at a temperature of 23 °C and relative humidity of 55% - 60%. One advantage of SSH libraries is that they are small enough to be printed multiple times on a single slide, thus increasing the accuracy of hybridization results. PCR products from 10 plant genes are also included on the slide for quality control purposes.

ii) Post-Processing of Slides: Slides are numbered and the boundaries of the array are marked with a diamond-tipped pen. Slides are allowed to dry for 24-48 hrs at room temperature. DNA is the cross-linked to the aminosilane matrix in a Stratalinker. Slides are then blocked by incubating twice in 0.1% SDS and washing in double-distilled water. Finally, DNA is denatured by boiling in double-distilled water. Slides are dried and stored at room temperature in a dessicator cabinet. Arrays are stable for at least 6 months when stored this way.

iii) Labelling and Hybridization RNA: TRI-REAGENT (Molecular Research Inc., Cincinnati, OH) is used to generate total RNA from cell lines or tissue samples. The manufacturer's protocol is followed, except that an extra phenol/chlorform extraction is added prior to the ethanol precipitation step, and RNA is resuspended to a final concentration of 6 mg/ml - 10 mg/ml. Targets were labelled using the FairPlay™ labeling Kit (Stratagene, La Jolla, CA). In this protocol, amino-allyl modified cDNA is generated in an oligo-dT primed reverse transcription reaction by replacing some of the dTTP in the recation mix with amino-allyl dUTP. Cy3- or Cy5 NHS ester (Amersham Pharmacia Biotech, Piscataway, NJ) is then covalently coupled to the amine residue. The labels are alternated, so that in some experiments the control sample is labeled with Cy3 and in others it is labeled with Cy5. The reactions are quenched, and the Cy3- and Cy5 labeled probes are combined and passed over a spin column (#A100610, Genisphere, Montvale, NJ) to remove unincorporated fluor. The probe is ethanol precipitated and resuspended in 30 μl hybridization buffer (0.57 mg/ml COT-1 DNA, 0.57 mg/ml poly-dA$_{40-60}$, 0.23 mg/ml yeast tRNA, 3.5X SSC, 0.3% SDS and 0.2 mg/ml BSA). The probe is then boiled for 2 min and then microfuged for 5 min at 4 °C. Supernatant is applied to slides and covered with a coverslip. Slides are incubated in a humidified chamber at 58 °C 12-16 hrs in the dark. Slides are washed in the following buffers at room temperature: 1) 10' in 2X SSC + 0.2% SDS; 2) 5' in 1X SSC + 0.2% SDS; 3) 1' in 2X SSC; and 4) 1' in 0.05X SSC. Slides are then dried by centrifuging at room temperature and scanned immediately. The printed matrix of PCR products on a single slide that has been hybridized with fluorescent-labeled control (Cy3, green) and experimental (Cy5, red) is shown in Figure 5. Note the faithfulness of the printing pattern and the reproducibility of spot size.

iv) RNA Isolation and Fluorescent Labeling using <u>Ultra-Small Samples</u>: In the past, using the standard Cy5 and Cy3 labeling protocol outlined above, 50 to 100 μg of total RNA was required to perform microarray analysis. Thus, gene expression profiling was limited to samples derived from either large tissues or cultured cell lines. A major role of the Biotechnology Core is to work with investigators to solve technical problems of this type. Recently, a new technology developed by NEN Life Sciences has reduced the amount of total RNA required for accurate gene expression profiling experiments down to 1 to 2 μg. This approach is based on Tyramide Signal Amplification, or TSA™, which adds considerable sensitivity through the addition of an enzymatic (horseradish peroxidase) amplification step. This is an important development for investigators working with very small amounts of total RNA (e.g. isolated cancer cells using techniques such as laser dissection). In the TSA™ approach (Figure 6), the RNA in each pool (control or experimental) is converted into either biotin-cDNA or dinitrophenyl (DNP)-cDNA using biotin-dCTP or DNP-dCTP and reverse transcriptase, according to the manufacturer's recommended protocol. These cDNAs are then mixed and placed in the center of the microarrays. A coverslip (Hybrislips, Corning) is carefully placed over the droplet and the samples are hybridized to microarrays. The slide is then hybridized as described above. Following incubation, microarrays are immersed in 30 ml of a wash solution (0.5X SSC + 0.01% SDS) in a 50 ml conical tube. Washes are at room temperature as follows: 1) 0.5X SSC + 0.01% SDS for 5 min; 2) 0.06X SSC, 0.01% SDS for 5 min; 3) 0.06X SSC for 2 min. Slides are then subjected to the TSA™ detection protocol (NEN Life Sciences,

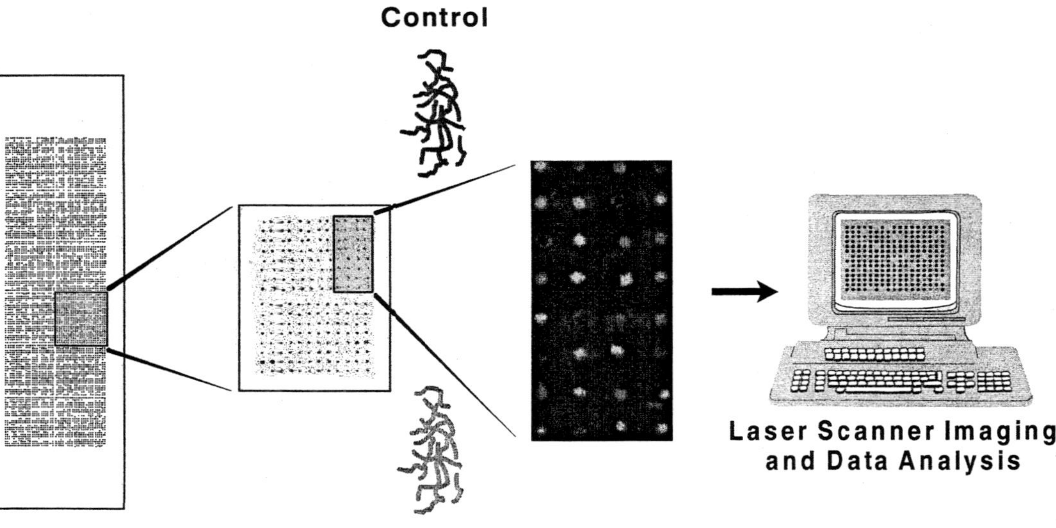

Figure 5. *Example of template of PCR products from SSH library that were printed onto a poly-L-lysine coated glass microscope slide. The excerpted panel shows the printing format for our robot. The center-to-center distance between individual DNA spots is approximately 200-250 µM and the average spot diameter is between 75 and 100 µM. These printed slides are then used for microarray experiments by exposing them to control and test cDNA (made from total RNA) that were differentially labelled with Cy3 (green) and Cy5 (red) fluorescent markers. The differential hybridization pattern is detected with a laser scanner and analyzed with special software.*

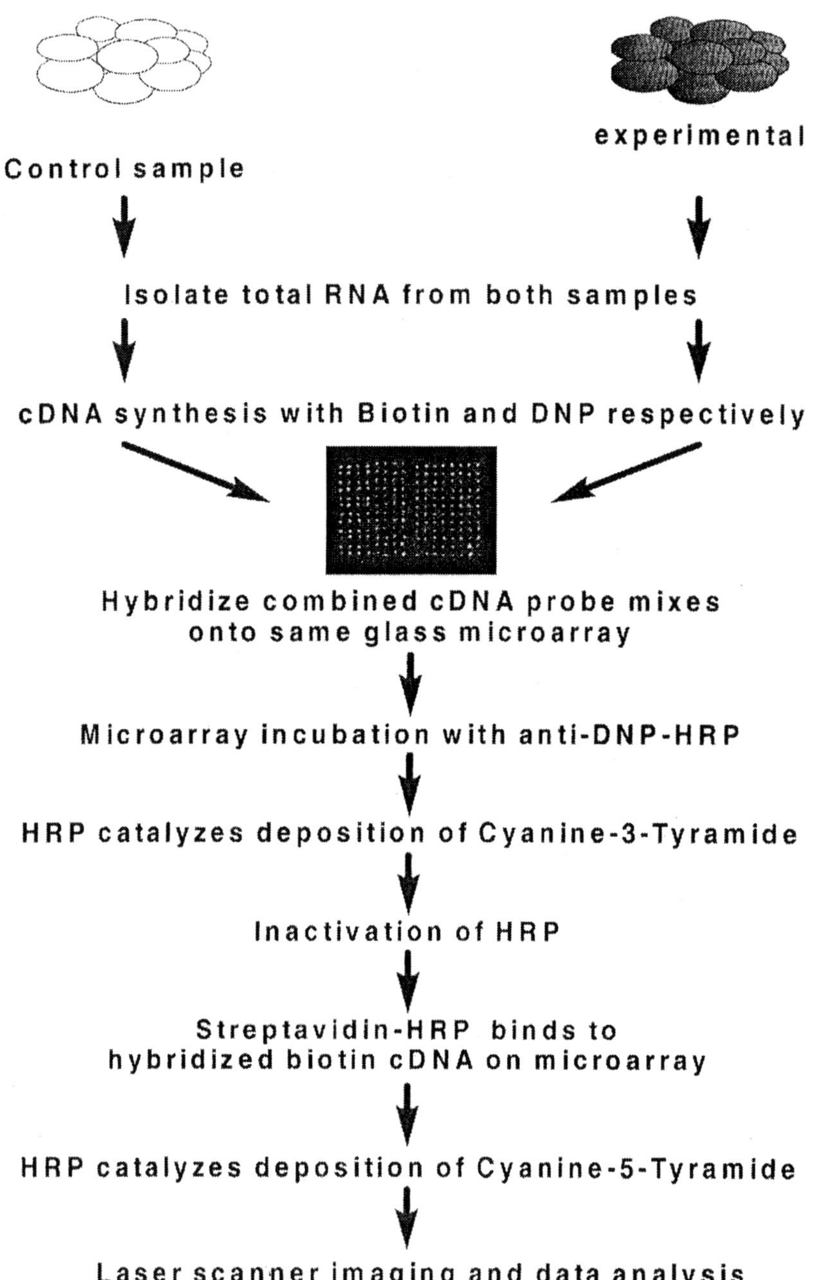

Figure 6. *Labeling of target RNA using Tyramide Signal Amplification approach.*

Boston MA), which sequentially tags the DNP-labeled cDNAs and the biotin-labeled cDNAs with the fluorescent reporters Cyanine-3-tyramide and Cyanine-5-tyramide, respectively. This protocol includes blocking the microarray in a buffer containing 10% goat serum, followed by incubation with the anti-DNP-HRP conjugate solution, and the Cyanine-3-tyramide solution. This deposits Cyanine-3-tyramide on all DNP-labeled cDNA. The HRP is then inactivated by incubating the microarrays in a solution containing sodium azide. The microarray is then incubated with a streptavidin-HRP conjugate solution, followed by the Cyanine-5-tyramide solution. This deposits Cyanine-5-tyramide on all biotin-labeled cDNA. Note that three 1 minute rinsing steps in sterile-filtered TNT buffer (0.1 M Tris-HCl, pH 7.5, 0.15 M NaCl, and 0.05% Tween-20) are included between each of the above incubations. The array is then rinsed for 1 minute in 30 ml of 0.06X SSC and dried by centrifuging at room temperature. A detailed description of TSA technology and the exact labeling and hybridization protocols are available online at (http://www.nenlifesci.com/pdf/MMMANUALPDF.pdf). Dried arrays are scanned as soon as possible after completion.

v) Validation Standards. An important control procedure for microarrays is to perform tests in which standards are used to examine the efficiency of hybridization and probe labelling, as well as the faithfulness of the hybridization signal to accurately reveal differential gene expression. Ten genes from the plant *Arabidopsis thaliana* (cat #252010, Stratagene, La Jolla, CA) are used as controls. These genes were selected because they do not match any known mammalian genes or ESTs in GenBank. PCR products of approximately 500 bp for each gene are placed into each plate of the library so that they were scattered throughout the array. Probes generated from mRNA corresponding to the spotted region of each Arabidopsis gene did not hybridize to any of the PC12-derived spots on the array. Three Arabidopsis genes were also spotted at the end of the array to serve as a reference for color balance during scanning. The mRNA for these three genes was spiked in equal amounts into both Cy3 and Cy5 labeling reactions. The remainder of the Arabidopsis mRNAs were spiked into each reaction at different known ratios. Results from on such control experiment are shown in Figure 7.

i) Data Collection: Because probes from both the test sample and the control sample are hybridized to the array simultaneously, the hybridization conditions are competitive. The relative abundance of a given target in the test population will therefore be reflected in the ratio of the fluorescence of the test target to that of the control target, and the results will be independent of the actual amount of DNA in each spot on the array. The spots containing Arabidopsis DNA serve as hybridization controls. Because an equal amount of three of the Arabidopsis RNAs is added to each target labeling reaction, fluorescence at these spots can be defined as a 1:1 ratio and used to correct for

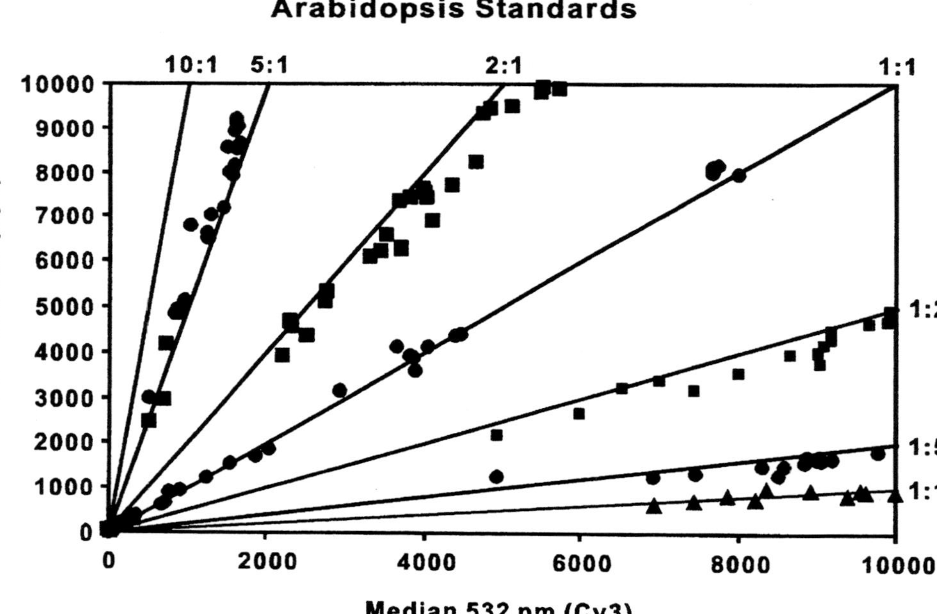

Figure 7. Ten genes from *Arabidopsis thaliana* were included in the microarray. Different ratios (10:1 to 1:10) and amounts of Arabadopsis mRNA were spiked into the labeling mix. The predicted results are represented by the solid lines and the symbols show the actual results. In is clear that the competitive hybridization for Cy3 and Cy5 labelled cDNA performed close to the predicted outcome during microarray experimentation.

any differences in labeling efficiencies between the Cy3 and Cy5 reactions. Fluorescence is measured with a Gene Pix 4000A microarray scanner (Axon Instruments, Palo Alto, CA). This instrument is a dual-wavelength scanner that scans simultaneously at 532 nm (green fluorescence) and 635 nm (red fluorescence), producing a single image and eliminating the need for image alignment. A representative image from a recent experiment in our laboratory is shown in Figure 8. In this particular experiment, control RNA was labelled with Cy5 and the experimental RNA was labelled with Cy3. Following normalization and background subtraction (which the software automatically performs), the ratios of the fluorescence for each spot are calculated. Targets in which the relative difference is at least 2-fold are selected for further analysis. The up- or down-regulation of each selected target gene is confirmed by real-time RT-PCR. Following analysis of the results from a microarray experiment, data can be displayed in a number of different formats

ix. Real-Time PCR. Once novel candidate genes and gene profiles have been identified, it is important to verify gene regulation by another approach. Real-time PCR was performed on RNA isolated from PC12 cells exposed to normoxia or hypoxia, as described in Methods. Total RNA is isolated as described above. RNA is then treated with Amplification Grade DNase (Life Technologies, Gaithersburg, MD) according to the manufacturer's protocol. RNA is cleaned by acid phenol:chloroform extraction followed by ethanol precipitation. cDNA is generated from 1-10 µg total RNA using the Superscript^{TM} First-Strand Synthesis Kit for RT-PCR (Life Technologies, Gaithersburg, MD) in a 40 µl reaction volume with oligo dT as primer. cDNA is then diluted to a working concentration (usually 0.1 µg/µl RNA equivalent).

Primers for real-time PCR are designed according to criteria outlined by Bustin (2000). Real-time PCR is performed using the SYBR Green I dye intercalation method (LightCycler – DNA Master SYBR Green I kit, Roche Diagnostics Corp., Indianapolis, IN) in a Smart Cycler® DNA amplification/detection system (Cepheid, Sunnyvale, CA). Fluorescence is detected during the 72 ^{O}C extension step of each cycle. Using the 2nd derivative of the growth curve, a cycle threshold (C_t) is determined. This cycle marks the entry into the log-linear phase of the reaction. The greater the amount of starting material (i.e., cDNA for the gene of interest), the earlier the C_t. Thus, if a gene is upregulated by hypoxia, the C_t for the hypoxic sample will be earlier than the C_t for the normoxic sample. At the end of each run, melt curve analysis (60 ^{O}C – 95 ^{O}C) is performed to determine product purity. PCR products are also isolated and sequenced to verify their identity.

The Future

The last 5 years has witness monumental changes in the landscape for basic and clinical research. These changes have been driven by sequencing of the

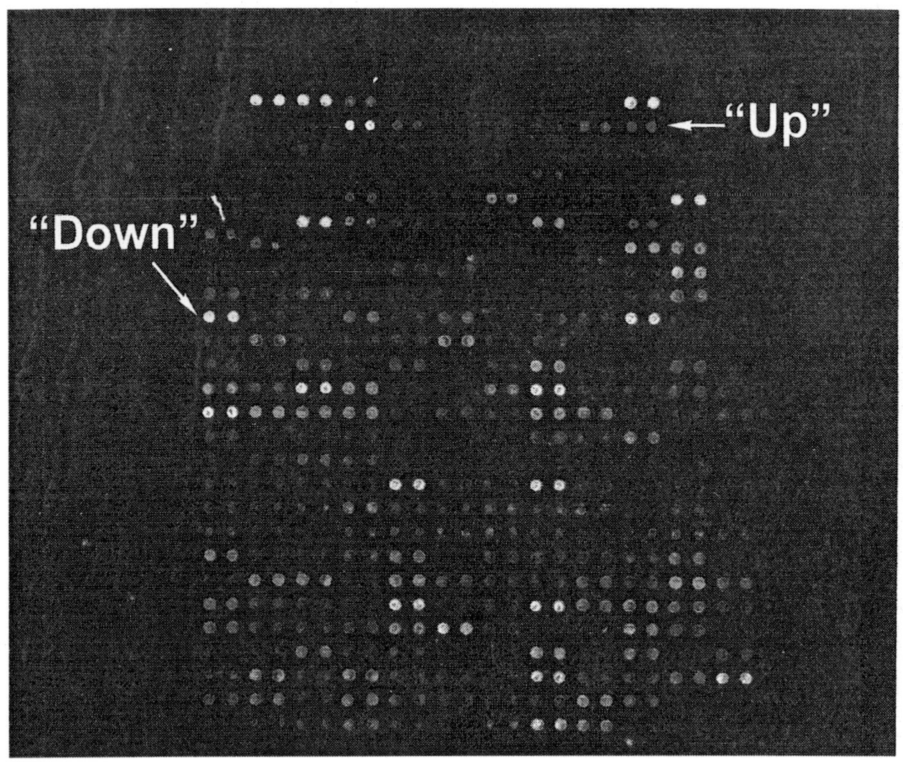

Figure 8. *Results from microarray experiment in which control is green and experimental is red. In this particular study two gene (SM-20 and MKP-1) to be regulated by hypoxia were identified. Another gene TH (tyrosine hydroxylase) which is known to be regulated by hypoxia in PC12 cells was a positive control for this experiment. The spots marked "up" and "down" are arabodopsis control standards that hybridize with spiked RNA .*

human and other genomes, and the development of new technologies for performing large-scale surveys of gene expression that are associated with complex physiological traits and disease processes. Nevertheless, there is a growing need for the refinement of these technologies and the development of new technologies that will enhance the sensitivity and efficiency for making global gene expression measurements. Many of the new approaches described in this book will provide investigators with more powerful approaches for more precise and efficient means for identifying and studying biologically significant gene expression profiles.

References

1. Schena, M.; Shalon, D.; Davis, R.W.; Brown, P.O. *Science* **1995**, 270, 467.
2. Bassett, D.E.; Eisen, M.B.; Boguski, M.S. *Nature Genetics* **1999**, 21, 51.
3. Brown, P.O.; Botstein, D. *Nature Genetics* **1999**, 21, 33.
4. Duggan, D.J.; Bittner, M.; Chen, Y.; Meltzer, P.; Trent, J. *Nature Genetics* **1999**, 21, 10.
5. Fambrough, D.; McClure, K.; Kazlauskas, A.; Lander, E.S. *Cell* **1999**, 97, 727.
6. Hill, C.S.; Treisman, R. (published online by the Science / AAAS at http://www.stke.org/cgi/content/full/OC_sigtrans;1999/3/pe1) **1999.**
7. Diatchenko, L.; Lay, Y.-F.; Campbell,, A.P. *Proc. Natl. Acad. Sci.* (USA) **1996**, 93, 6025.
8. Endege, W.O.; Steinmann, K.; Boardman, L.; Thibodeau, S.; Schlegel, R. *BioTechniques* **1999**, 26, 542.
9. Gurskaya, N.G.; Diatchenko, L.; Chenchik, A. *Anal. Biochem.* **1996**, 240, 90.
10. Hubank, M.; Schatz, D.G. *Nucleic Acid Res.* **1994**, 22, 5640.
11. Welford, S.M.; Gregg, J.; Chen, E.; Garrison, D.; Sorensen, P.; Denny, C.; Nelson, S. *Nucleic Acids Res.* **1998**, 15, 3059.
12. Kim, H.; Yuan, Y.; Seta, K.; Moore, T.; Beitner-Johnson, D.; Millhorn, D.E.*FASEB J.* **2000**, A75.
13. Nelson, S.F.; Denny, C. In: DNA Microarrays: A Practical Approach, edited by Mark Schena. Oxford University Press. New York, **1999**, 43.
14. Wong, B.; Park, C.; Lee, S.; Choi, Y. *CLONTECHniques* **1996**, 11, 32.
15. Beitner-Johnson, D.; Seta, K.; Yuan, Y.; Kobayashi, S.; Millhorn, D.E. *Parkinsonism and Related Disorders.* **2000**, 273.

Chapter 2

Properties, Manufacturing, and Applications of MicroArrays of Gel-Immobilized Compounds and Cells on a Chip

Andrei D. Mirzabekov[1]

Joint Biochip Program of Engelhardt Institute of Molecular Biology, Russian Academy of Sciences, 119991, Moscow, Russia
[1]Current address: Biochip Technology Center, Argonne National Laboratory, 9700 South Cass Avenue, Argonne, IL 60439–4833

A MicroArray of Gel-Immobilized Compounds and Cells on a chip (MAGICChip) contains oligonucleotides, DNA, proteins, or other compounds tethered within gel micropads. The gel pads can be used both as a support for immobilization and as nanoliter test tubes to carry out different chemical or enzymatic reactions with tethered compounds. Nucleic acid hybridization; specific interactions between nucleic acid, proteins, and small ligands; and kinetic and thermodynamic analysis of some of these interactions have been carried out by using MAGICChips. MAGICChips have also been applied to perform DNA fractionation, ligation, PCR amplifications, and minisequencing. Different processes were monitored within the microchip pads by using specially constructed fluorescence microscopes and laser scanners, as well as a MALDI mass spectrometer. The oligonucleotide microchips have been applied to *i)* identify different bacteria and viruses and different genes (toxin and drug resistance genes) in their genomes; *ii)* detect mutations and nucleotide polymorphism in the human genome, microorganisms, and in viruses; and *iii)* monitor gene expression. The protein microchips have been tested to determine their effectiveness in detecting specific antigens and antibodies, evaluate ligand specificity for receptors, and measure kinetics of enzymatic reactions and inhibition. The presence of different antibiotics was monitored with bacterial cell microchips.

Introduction

Biochip technology is becoming a well-established field in many companies and laboratories *(for review, see references 1,2)*. The development of oligonucleotide microchips was initiated at the Engelhardt Institute of Molecular Biology in 1988 to facilitate nucleic acid sequencing by a new method — sequencing by hybridization *(3)*. In 1989, polyacrylamide hydrogels were demonstrated to be a convenient support to immobilize oligonucleotides and subsequently produce a hybridization «microchip» *(4,5)*. Thereafter, a multidisciplinary team of molecular biologists, chemists, physicists, computer scientists, and engineers began to develop **MicroArray of Gel-Immobilized Compounds and Cells on a Chip** (MAGICChip) technology. The organization of a joint Biochip Program by Engelhardt Institute of Molecular Biology, Russian Academy of Sciences (EIMB) and Argonne National Laboratory (ANL), in 1994 significantly accelerated the advance of the MAGICChip technology and its applications. Studies of MAGICChip technology have been carried out in collaboration with a number of groups in the United States, Russia, and the European community and have been supported by grants from U.S. and Russian government agencies and U.S. companies.

MAGICChip Properties

The MAGICChip is a glass slide containing gel pads (Fig. 1A) attached by chemical bonds to a glass surface *(6)*. Different molecules, such as oligonucleotides, DNA fragments, proteins, and low-molecular-weight ligands, have been immobilized within the microchip gel-pads. The shape and size of the gel pads can vary. Typically, the chips contain polyacrylamide hydrogel pads measuring $100\times100\times20$ μm and 0.2 nanoliters in volume, and spaced 200 μm apart. The technology of manufacturing the microchips containing gel pads as small as $10\times10\times5$ μm has also been developed *(7)*. Other gels that are more porous *(8)* or stable than polyacrylamide, for example, copolymers of acrylamide, methacrylamide and other compounds, can be used in microchip manufacturing. The glass surface between the pads is hydrophobic, and so the pads are isolated from each other when a wet microchip is exposed to air or covered by mineral oil or other water-insoluble solvents. The gel pads containing water solution of different compounds do not communicate with each other under these conditions and behave like isolated nanoliter test tubes. Different chemical and enzymatic reactions can be carried out separately in such gel pad test tubes.

Tethering the molecules within a three-dimensional gel structure significantly (100–1,000 times) increases immobilization capacity, as compared with the glass surface. Interactions of macromolecules with gel-immobilized compounds occur in a more homogeneous water surrounding and they are more like water solution processes than heterophase interactions with glass-immobilized compounds. This advantage of gel-based microchips was supported by on-chip kinetic and thermodynamic measurements of specific interactions and reactions *(9,10)*. Five percent polyacrylamide gel pads of the microchip are accessible for penetration of single-stranded DNA of up to 150 nucleotides in length and proteins of up to 200,000 Da in molecular weight. More-porous gels were used for longer DNA of up to 400 nucleotides in length and larger proteins or their complexes *(8)*.

Oligonucleotide and some protein microchips can be used many times, stored for several months, and mailed.

Manufacturing

Synthetic or natural compounds are tethered to gel pads within their three-dimensional structure in the course of microchip manufacturing. Two different methods have been developed for this purpose.

The first method, blank microchip manufacturing, consists of four successive steps *(6)*. In the first step, the micromatrix of the gel pads (blank microchip) is produced by photoinduced polymerization of the layer of the monomer solution, usually acrylamide and bisacrylamide containing a photo-initiator, illuminated through a mask. In the second step, the gel pads of a blank microchip are then activated to give rise to reactive chemical groups, usually aldehyde or hydrazide, for chemical coupling of the oligonucleotides or DNA *(11)* to be immobilized. In the third step, solutions of different compounds are applied specifically to each gel pad by a robot. Finally, applied compounds are tethered to the gel within the pads by a covalent bond.

The second, copolymerization method, *(7),* consists of two steps. A solution of monomers containing a compound with an attached monomer group is first applied by a robot as a drop on a glass slide. Then, the copolymerization of free monomers with a monomer attached to the compound is induced by illumination with UV light or by a chemical

MAGICChip

A

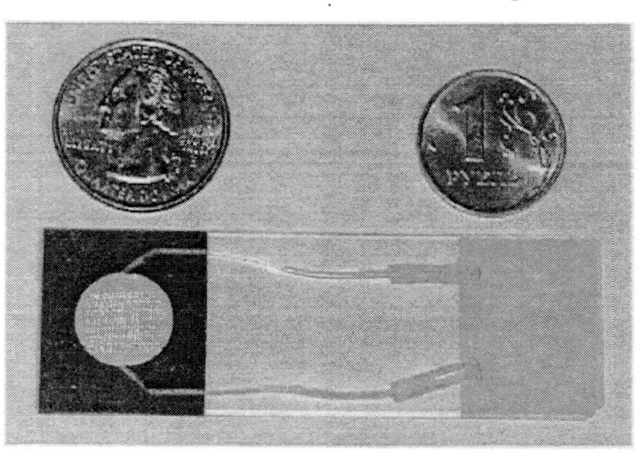

Biochip Analyzers

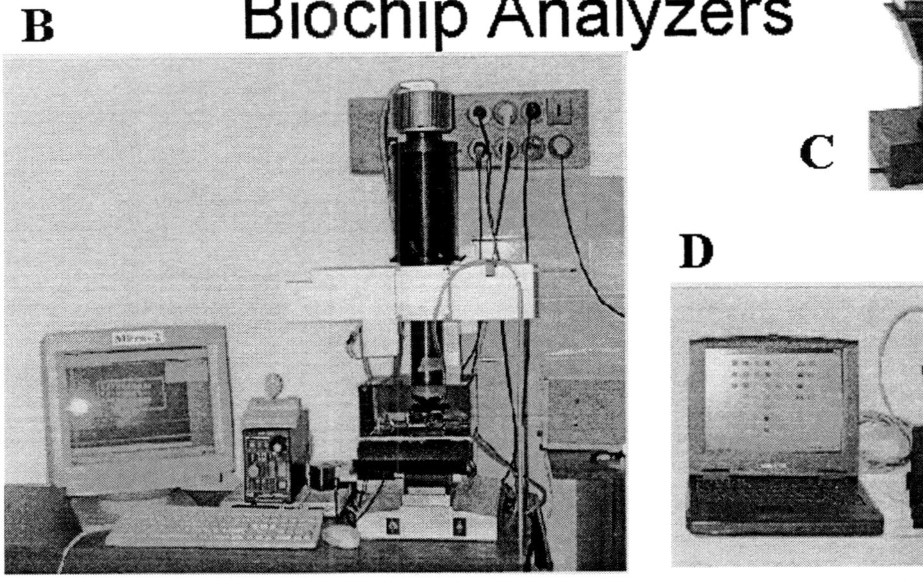

Figure 1. MAGICChip and microchip analyzers.

A. MAGICChip in a chamber to carry out hybridization and PCR amplification

B. Reseach fluorescence microscope

C, D. Portable fluorescence analyzer complete with Polaroid camera (C) or CCD-camera and lap-top computer (D).

initiator at room temperature. The monomer can be incorporated into oligonucleotides during their synthesis; incorporation of a monomer into DNA or protein molecules should be carried out as a separate chemical reaction. The copolymerization method is simpler, faster, and less expensive than blank microchip manufacturing and can be easily done in a research laboratory.

Three different robots have been tested for microchip production: a one-pin robot specially constructed by our group *(6)* and a piezoelectric dispenser (Packard) are suitable for both methods, and high-output four-pin robot (Affymetrix) and the piezoelectric dispenser are efficient for copolymer hydrogels microchip production.

Microchip Monitoring

Different methods have been tested for monitoring MAGICChips, and these methods are described in the following paragraphs.

Fluorescence Microchip Analyzer

A specially devised microchip analyzer (Fig. 1B) consists of a several-wavelength fluorescence microscope, a CCD camera, a regulated thermotable, and proper computer software *(6, 10, 12)*. The microscope is equipped with a wide-field, long-focus, high-aperture objective. This allows one to carry out quantitative parallel measurement in real time for all microchip gel pads in an area of 12 mm in diameter at one, two, or more wavelengths. The use of such an objective is also convenient to monitor the microchip directly in the reaction chamber so that the kinetics of the processes can be measured at precisely regulated or varied temperatures. A simple, inexpensive (less than \$5,000), large-field fluorescence microscope supplied with a Polaroid photocamera (Fig. 1C) or CCD-camera and a lap-top computer (Fig.1D) has been constructed. The microchip in this analyzer was illuminated from side by lasers or photodiods *(13, 14)*.

Laser Scanning

A portable, inexpensive laser scanner has been developed in our group (14) for analysis of microchips in a hybridisation chamber. A laser scans the microchip successively site after site and provides a high dynamic range and sensitivity of measurements.

Mass Spectrometry

A Matrix-Assisted Laser Desorption Ionization (MALDI) mass spectrometry was successfully applied to detect short oligonucleotides hybridized with oligonucleotide microchips *(15)*.

Microscopy

Enzymatic activities were detected by using colored reactions and observed by using a conventional microscope of low magnification *(8)*.

MAGICChips of Different Composition

Many natural and synthetic biopolymers, as well as small ligands, can be tethered within microchip gel pads by both the "blank microchip" and copolymerization methods.

Oligonucleotide and DNA Microchips

Chemically active attachment groups and monomers can be incorporated into oligonucleotides in the course of their synthesis *(16)*.

Two types of oligonucleotide microchips have been produced: custom-made and generic microchips. A custom-made microchip contains a set of specific oligonucleotides, whereas the generic microchip accumulates all possible oligonucleotides of a certain length — for example, 4,096 of hexadeoxyribonucleotides *(9, 17, 18)*. 6-mer duplexes are melted at rather low temperatures, which is often inconvenient for their use. Therefore, the hexamers in the generic microchips were extended from both ends by using a universal base, 5-nitroindole, or by using the mixture of four nucleotides (A, T, G, and C) to substitute the formation of 6-base-pair duplexes with those of more stable 8-base-pair ones (*10*). In addition, modified nucleotides can be incorporated into 6-mer to stabilize their duplexes (*19*).

Natural DNA molecules should be modified to introduce specific groups needed for DNA attachment to the gel structure *(20)*. Polymerase chain reaction-amplified DNA can be prepared by using synthetic oligonucleotides containing such groups as primers *(20)*.

Protein Microchips

Protein molecules contain active groups, such as amino- and sulfhydride groups, that can be used for immobilization (8) or attachment of copolymerization monomer residues. Specific amino acid residues containing these groups or an oligohistidine chain can be incorporated into genetically engineered proteins at a specified site. As an alternative, the fixation of protein molecules inside polyacrylamide gel was carried out by using glutaraldehyde *(8)*.

Polysaccharide Microchips

Polysaccharides can be activated for their attachment to the gel matrix or for incorporation into co-monomer pendant groups. For example, polysaccharide tails of antibodies were activated by producing aldehyde groups by oxidation of some monocharide residues with sodium periodate *(8)* and attached to a gel matrix similar to that resulting from the immobilization of aldehyde-containing oligonucleotides and DNA *(6, 11)*. This method could be used for manufacturing polysaccharide microchips.

Small Ligand Microchips

The biotin group, as well as some Ni-chelate groups, has been incorporated into a microchip gel matrix by copolymerization to specifically bind avidine or genetically engineered proteins containing the oligo-histidine group, respectively.

Cell microchips

Different strains of *E. coli,* sensitive or resistant to various antibiotics, have been immobilized within polyacrylamide gel pads. Such cell microchips have been applied for detection of these antibiotics *(21)*.

Specific Interactions on MAGICChips

Hybridization

The hybridization of nucleic acid on biochips is well-developed technology. MAGICChips based on gel-immobilized oligonucleotides or DNA provide high efficiency in hybridization. The theory of the hybridization of gel-immobilized oligonucleotides was developed *(21)*. Reliable discrimination of perfect duplexes formed by 15-20 mers with DNA or RNA from duplexes containing a single mismatch at the central position can usually be achieved *(10, 23)*. The hybridization with oligonucleotide microchips can be used for fractionation of DNA; the fractions can be subsequently analyzed by hybridization with another sequencing microchip *(24)*.

Equilibrium melting curves of duplexes formed on oligonucleotide microchips were measured by using the fluorescence microchip analyzer *(10,19)*. Melting curves data were used to calculate melting temperatures (T_m) and thermodynamic parameters of duplexes, entropy (ΔS), enthalpy (ΔH), and free energy $(\Delta G^{o)}$ of the hybridization. The use of the generic microchips enables one to measure, in one experiment, the thermodynamics of the hybridization for the perfect duplex and mismatched duplexes containing all possible mispairings in all positions of the 6-mer duplexes (25). This efficient approach has also been used to estimate the effect of incorporating modified nucleotides into oligonucleotides on their duplex stability and on the efficiency of discriminating between perfect and mismatched duplexes (19).

Contiguous Stacking Hybridization

The shorter the duplexes, the higher the discrimination between perfect and mismatched ones, and at the same time, the lower their stability. The low stability of short duplexes can be increased by contiguous stacking hybridization. For example, the melting temperature of a 5-mer duplex can be increased by 5–10°C if it is stacked to an 8-mer duplex and forms a contiguous 13-mer duplex containing a broken phosphodiester bond in one strand (5, 26). The use of contiguous stacking interaction also allows one to increase the efficiency of, for example, an 8-mer generic microchip to the level approaching that of a 13-mer microchip. The stabilization effect of all pairs of stacked bases in deoxyribonucleotides has been measured with oligonucleotide microchips (27).

Protein-Protein Interactions

MAGICChips containing immobilized polyclonal and monoclonal antibodies and antigens were used for their mutual identification *(8)*. This opens up a possibility to develop monoclonal antibody microchips that could be used to determine the presence of different proteins in different cells and tissues. These monoclonal antibody microchips could complement the use of gene expression microchips. One can consider the application of protein microchips for the detection of specific interactions of other proteins, such as regulatory factors and receptors.

Interactions of DNA and Proteins with Each Other and with Small Ligands

Specificity of the interactions of the DNA-binding drug Hoechst 33258 (bisbenzimide, Hoechst, Germany) with a large spectrum of DNA duplexes was studied by using generic oligonucleotide microchips. Free energy of the interaction was measured for some of these duplexes *(9)*. The specificity of interaction of bacterial histone-like protein HU with single- and double-stranded duplexes was also evaluated by using the 6-mer generic microchip *(28)*.

Specific Reactions on MAGICChips

Chemical or enzymatic reactions can be carried out simultaneously with all microchips gel pads or in each gel pad separately. In the first case, the microchip is covered with reaction solution, and reactants diffuse into all gel pads *(24)*. In the second case, wet gel pads are separated from each other by air *(6)* or water-insoluble solvents *(7, 29)*, such as chloroform and mineral oil. Using a hydrophobic glass surface to space the pads prevents any exchange of water solutions between the pads. Reactants can diffuse into all gel pads from the chloroform or a water layer placed on the chloroform (7). Specific reactants can be applied to each pad by using a pin robot or a piezoelectric dispenser.

Chemical Reactions

Chemically reactive groups, such as aldehyde-, amino-, hydrazide- can be introduced into the gel matrix by using a specific gel treatment *(11, 20)* or copolymerization procedure *(7)*. Different oligonucleotides or DNAs containing complementary chemical groups (amino- or aldehyde groups) were applied to each gel pad by robot, and the coupling reaction between the gel and these compounds occured in air *(6)*. In some cases, to stabilize and reduce aldamine bonds formed between gel and oligonucleotides, the microchips were covered with chloroform containing a reducing agent (7). The agent diffused into the water phase of the pads and stabilized the covalent bonds between the gel and oligonucleotide molecules.

Enzymatic Reactions

A number of enzymatic procedures were carried out within microchip gel pads containing tethered oligonucleotide and covered with a buffer solution or with mineral oil: phosphorylation, ligation *(24, 30)*, DNA polymerase reactions for single-base extension *(31)*, primer chain extension, and PCR amplification *(29, 32)*.

Enzymes are known to be stabilized by immobilization within a gel, and therefore MAGICChip with tethered enzymes could be a convenient analytical instrument. Different colored reactions were carried out with a number of microchip immobilized enzymes *(8)*. Dephosphorylation of substrate-yielding fluorescent precipitate with gel-tethered alkaline phosphatase and the enzyme inhibition with phenylalanine was measured by using the fluorescent microchip analyzer. The kinetic data obtained for the immobilized enzyme within the gel were close to those measured in solution *(8)*.

Biomedical Applications of MAGICChips

Oligonucleotide biochips have been tested in a number of applications: for analysis of mutations and single nucleotide polymorphism, for *de novo* sequencing by hybridization and sequence proofreading, for identification of pathogenic and non-pathogenic microorganisms and viruses, for detection of toxin and drug resistance genes, and for analysis of gene expression in leukemias.

Mutation and Single Nucleotide Polymorphism Analysis

Several technologies and methods have been developed and tested for analysis of mutations and single nucleotide polymorphisms (SNP) in a number of genes in different organisms.

Hybridization Analysis.

The most simple and advanced procedure of mutation and SNP analysis is based on PCR amplification of DNA in solution and the hybridization of this DNA with a set of mutation-specific microchip oligonucleotides. This scheme was shown in several cases to reliably identify the 30 most frequently occurring rifampicin-resistant TB mutations — both in cell cultures as well as in patient sputum (30). Fluorescence labeling of DNA samples and using a fluorescent analyzer to monitor the microchip hybridization is the most convenient analytical approach. Manufacturing inexpensive microchips and using a simple, but efficient, fluorescent microchip analyzer developed by our group (13, 14) will make analysis available to many research and hospital laboratories. The use of contiguous stacking hybridization and analysis of unlabeled DNA with a MALDI mass spectrometer was also tested (15), and it could be a promising diagnostic method for future use.

A similar method involving hybridization of PCR-amplified DNA by using an oligonucleotide microchip was also applied to analyze β-thalassemia mutations in β-globin gene *(23)*; SNP in mu-opioid receptor *(33),* which is believed is involved in drug addiction; SNP in smallpox viruses of human and various animal origins *(34)*, and) to identify HLA DQH1 human histocompatibility gene alleles *(35).*

Mutation Analysis by Proofreading with Generic Microchips.

The sequencing by hybridization is based on the hybridization of DNA with the whole set of oligonucleotides of a certain length (3). We have proposed (4) and developed sequencing and sequence proofreading by hybridization of DNA with a generic microchip. The generic microchip containing all 4,096 6-mers has been manufactured (17, 18) and used for sequencing short (up to 80 nucleotides long) DNA that does not contain any long repeats (18). The generic microchip could be a convenient instrument for resequencing and for detecting of known and new mutations in the DNA of known structure. This microchip was applied to detect mutations in oral poliovirus vaccine (17). The generic microchip was demonstrated to be useful for sequence analysis of complex mixtures of DNA fragments of V-D-J genes of CDR3 T-all receptors (36).

Single-Base Extension.

Another approach for SNP and mutation analysis is the combination of PCR amplification in a tube with on-chip single-base extension. Single-base-extension analysis of β-thalassemia mutations was carried out with thermostable DNA polymerase at elevated temperature, which enhanced the extension signals by about 10 times (31). Two variations of the method have been developed: multiprimer and multibase extensions that allowed us to identify four nucleotide polymorphism and achieve about 10 times difference for correct and false positive signals.

On-Chip PCR Amplification.

On-chip amplification was performed in two variations. The first approach is over/on-chip amplification (32). The PCR reaction was carried out in parallel in solution over the microchip with both unbound primers and on the microchip within the gel pads containing one immobilized and one soluble diffused primer. The method was efficient for simultaneous detection of different rifampicin-resistant mutations in tuberculosis (TB). It is inexpensive, sufficiently sensitive to analyze about 1000 cells, and takes about 1.5 h instead of the several weeks typically needed for bacteriological tests. The method was also applied to identify a single nucleotide polymorphism that differentiates shiga- and shiga-like toxin genes.

In the second, only on-chip PCR amplification method *(29)*, the DNA polymerase reactions with oligonucleotide primers take place in individual gel pads separated from each other by mineral oil. All reaction components (including enzyme and amplified DNA) diffuse into the pads. Specific sets of oligonucleotide primers that can be used for a number of successive amplifications are immobilized within the gel pads. PCR reaction is inhibited if both, forward and reverse primers are tethered to the gel. Therefore, the structure of the primers is determined in such a way that they can be detached from the gel by specific enzymatic treatment; the reaction is performed under oil, and therefore released primers remain within a gel pad. Inactive additional primers can be immobilized within a pad and activated to be used in the second round of PCR amplification, allele-specific primer extension, or single-base extension. This approach has been used to identify five mutations, which are localized at different genes of the *M. tuberculosis* genome, which determine the resistance of these pathogenic microorganisms to rifampicin, streptomycin, and isoniazid *(29)*.

These on/over- and on-chip amplifications were tested in preliminary studies for analysis of SNP in the human genome.

Identification of Toxin- and Drug-Resistant Genes

These genes can be determined by using different techniques described for the detection of mutations and SNP, such as the hybridization of labeled PCR-amplified DNA with oligonucleotide microchip, on-chip single-base extension and on-chip PCR amplification.

Anthrax toxins, shiga, and shiga-like toxin genes, as well as ampicillin-resistant plasmid genes, were detected by over/on-chip PCR amplification *(32)*.

Identification of Bacteria and Viruses

16S ribosomal RNA is highly conservative in evolution, but at the same time, it is sufficiently divergent to be used for identification of microorganisms. Oligonucleotide microchips have been manufactured that contain oligonucleotides complementary to specific regions of 16S RNA of various microorganisms. Such microchips have been used successfully to analyze nitrifying bacteria *(37)* and to determine the types of bacteria present in Siberian high-temperature petroleum reservoirs *(38)*. This method was applied to identify *B. anthracis* and discriminate it from very similar B. *cereus, B. thuringiensis, B. medusa, B. mycoides*, and *B. subtilis (13, 39)*. The method was sufficiently reliable in these experiments to discriminate closely related bacteria that are different only in one base in 16S RNA.

A simple procedure has been developed in conjunction with this method to isolate RNA from bacterial lysate, partially degrade it to the size needed for hybridization, and fluorescently label it. All of these steps are performed by applying the bacterial lysate on a column and washing the column successively with different reagents *(13)*.

Identification of viruses can be carried out by methods used to detect any of its genes by the methods described above. Different strains of smallpox belonging to humans and different animals were identified and discriminated from each other by using the hybridization of viral-amplified DNA with specific microchip oligonucleotides *(34)*.

Gene Expression Analysis

Microchips for expression analysis can be manufactured by tethering either proper oligonucleotides or cDNA *(19)* or by simultaneously tethering both oligonucleotides and cDNA on different gel pads of the same microchips (V. Shick, unpublished results).

Oligonucleotide microchips have been applied to identify spliced mRNA in leukemia cells caused by chromosomal rearrangement *(40)*. This approach could be a reliable method for diagnosing various leukemias and lymphomas.

Acknowledgements

I would like to thank all of my colleagues, who are authors of the papers cited, for stimulating discussions.

This work was sponsored by the U.S. Department of Energy, under Contract No. W-31-109-ENG-38 and by grant 5/2000 from the Russian Human Genome Program.

References

1. Reviews: The Chipping Forecast. *Nature Genetics* **1999**, *21*, 1–60.
2. Tillib, S., and Mirzabekov, A. Advance in the analysis of DNA sequence variations using oligonucleotide microchip technology. *Current Opin. Biotech.*, **2001**, *12(1)*, 53-58.
3. Lysov, Yu.; Florentiev, V.; Khorlin, A.; Khrapko, K.; Shick, V.; Mirzabekov, A. A new method to determine the nucleotide sequence by hybridizing DNA with oligonucleotides. *Proc. Acad. Sci. USSR (English Ed.)* **1988**, *303*, 436– 438.
4. Khrapko, K.; Lysov, Yu.; Khorlin, A.; Shick, V.; Florentiev, V.; Mirzabekov, A. An oligonucleotide hybridization approach to DNA sequencing. *FEBS Letters* **1989**, *256*, 118–122.

5. Khrapko, K.; Lysov, Yu.; Khorlin, A.; Ivanov, I.; Yershov, G.; Vasilenko, S.; Florentiev, V.; Mirzabekov, A. A method for DNA sequencing by hybridization with oligonucleotide matrix. *DNA Sequence* **1991,** *1,* 375–388.

6. Yershov, G.; Barsky, V.; Belgovskiy, A.; Kirillov, E.; Kreindlin, E.; Ivanov, I.; Parinov, S.; Guschin, D.; Drobyshev, 7. A.; Dubiley, S.; Mirzabekov, A. DNA analysis and diagnostics on oligonucleotide microchips. *Proc. Natl. Acad. Sci. USA.* **1996,** *93,* 4913–4918.

7. Vasiliskov, A.V.; Timofeev E.N.; Surzhikov, S.A.; Drobyshev, A.L.; Shick, V.V.; Mirzabekov, A.D. Fabrication of microarray of gel-immobilized compounds on a chip by copolymerization. *BioTechniques.* **1999,** *27,* 592–606.

8. Arenkov, P.; Kukhtin, A.; Gemmell, A.; Chupeeva, V.; Mirzabekov, A. Protein microchips: Use for immunoassay and enzymatic reactions. *Anal. Biochem.* **2000,** *278,* 123–131.

9. Drobyshev, A.; Zasedatelev, A.; Yershov, G.; Mirzabekov, A. Massive parallel analysis of DNA-Hoechst 33258 binding specificity with a generic oligodeoxyribonucleotide microchip. *Nucl. Acids Res.* **1999,** *27,* 4100–4105.

10. Fotin, A.; Drobyshev, A.; Proudnikov, D.; Perov, A.; Mirzabekov, A. Parallel thermodynamic analysis of duplexes on oligodeoxyribonucleotide microchips. *Nucl. Acids Res.* **1998,** *26,* 1515-1521.

11. Timofeev, E.; Kochetkova, S.; Mirzabekov, A.; Florentiev, V. Regioselective immobilization of short oligonucleotides to acrylic copolymer gels. *Nucl. Acids Res.* **1996,** *24,* 3142–3148.

12. Barskii, Ya.; Grammatin, A.; Ivanov, A.; Kreindlin, E.; Kotova, E.; Barskii, V.; Mirzabekov, A. Wide-field luminescence microscopes for analyzing biological microchips. *J. Opt. Technol.* **1997,** *65,* 938–941.

13. Bavykin, S.G.; Akowski, J.P.; Zakhariev, V.M.; Barsky, V.E.; Perov, A.N.; Mirzabekov, A.D. Portable system for microbial sample preparation and oligonucleotide microarray analysis. **2001.** *Appl. Environ. Microbiol.* 67, 922-928.

14. Barsky, V.; Perov, A.; Tokalov, S.; Chudinov, A.;.Kreindlin, E.; Sharonov, A; Kotova, E.; Mirzabekov, A. Biochips as arrays of biosensors: equipment, fluorescent labels, and applications. **2001** (*unpublished*).

15. Stomakhin, A.; Vasiliskov, V.; Timofeev E.; Schulga, D.; R. Cotter, A.; Mirzabekov. DNA sequence analysis by hybridization with oligonucleotide microchips: MALDI mass spectrometry identification of 5-mers contiguously stacked to microchip oligonucleotides. **2000.** *Nucl. Acids Res.* **28** (5), 1193-1198

16. Surzhikov, S.; Timofeev, E.; Chernov, B.; Golova Yu.; Mirzabekov A. Advanced method for oligonucleotide deprotection. **2000.** *Nucl. Acids Res.* **28** (8), e29

17. Proudnikov D.; Kirillov E.; Chumakov K.; Donlon J.; Rezapkin G.; Mirzabekov, A. Analysis of mutations in oral poliovirus vaccine by hybridization with generic oligonucleotide microchips. *Biologicals* **2000**, *28*, 57-66.

18. Chechetkin, V.R.; Turygin, A.Y.; Proudnikov, D.Y.; Prokopenko, D.V.; Kirillov, E.V.; Mirzabekov, A.D. Sequencing by hybridization with the generic 6-mer oligonucleotide microarray: an advanced scheme for data processing. *J. Biomol. Struct. Dyn.* **2000**, *18*, 83-101.

19. Timofeev, E.; Mirzabekov, A. Binding specificity and stability of duplexes formed by modified oligonucleotides with a 4,096-hexanucleotide microarray. *Nucl. Acids Res.* **2001**, *29*, 2626-2634.

20. Proudnikov, D.; Timofeev, E.; Mirzabekov, A. Immobilization of DNA in polyacrylamide gel for the manufacture of DNA and DNA-oligonucleotide microchips. *Anal. Biochem.,* **1998**, *259*, 34–41.

21. Fesenko, D.; Nasedkina, T.; Mirzabekov, A. Bacterial microchip: proof of principle by testing antibiotic resistance in bacteria. *Proc. Russ. Acad. Sci.* **2001**, (*in press*).

22. Livshits, M.; Mirzabekov, A. Theoretical analysis of the kinetics of DNA hybridization with gel-immobilized oligonucleotides. *Biophys. J.* **1996**, *71*, 2795–2801.

23. Drobyshev, A.; Mologina, N.; Shick, V.; Pobedimskaya, D.; Yershov, G.; Mirzabekov, A. Sequence analysis by hybridization with oligonucleotide microchip: identification of beta-thalassemia mutations. *Gene.* **1997**, *188*, 45–52.

24. Dubiley, S.; Kirillov, E.; Lysov, Yu.; Mirzabekov, A. Fractionation, phosphorylation, and ligation on oligonucleotide microchips to enhance sequencing by hybridization. *Nucl. Acids Res.* **1997**, *25*, 2259–2265.

25. Khomyakova, E., Livshits, M.; Sharonov, A.; Prokopenko, D.; Mirzabekov, A. Parallel analysis of perfect and mismatched DNA duplexes of various GC contents with generic hexamer nucleotides. **2001** (*unpublished*).

26. Parinov, S.; Barsky, V.; Yershov, G.; Kirillov, E.; Timofeev, E.; Belgovskiy, A.; Mirzabekov, A. DNA sequencing by hybridization to microchip octa- and decanucleotides extended by stacked pentanucleotides. *Nucl. Acids Res.* **1996**, *24*, 2998–3004.

27. Vasiliskov, V.; Prokopenko, D.; Mirzabekov, A. Parallel multiplex thermodynamic analysis of coaxial base stacking in DNA duplexes by oligodeoxyribonucleotide microchips. *Nucl. Acids Res.* **2001**, *29*, 2303-2313.

28. Krylov, A.; Zasedateleva, O.; Prokopenko D.; Rouviere-Yaniv, J.; Mirzabekov A. Massive parallel analysis of the binding specificity of histone-like protein HU to single- and double-stranded DNA with generic oligodeoxyribonucleotide microchips. *Nucl. Acids Res.* **2001**, *29*, 2654-2660.

29. Tillib, S.V.; Strizhkov, B. N.; Mirzabekov, A.D. Integration of multiple PCR amplifications and DNA mutation analyses by using oligonucleotide microchip. *Anal. Biochem.* **2001**, *292*, 155-160.

30. Mikhailovich, V.; Lapa, S.; Gryadunov, D.; Sobolev, A.; Strizhkov, B.; Chernyh, N.; Skotnikova, O.; Irtuganova, O.; Moroz, A.; Litvinov, V.; Vladimirsky, M.; Perelman, M.; Chernousovsa, L.; Erokhin, V.; Zasedatelev, A.; Mirzabekov, A. Identification of rifampicin-resistant *Mycobacterium tuberculosis* strains by hybridization, PCR, and ligase-detection reaction on oligonucleotide microchips. *J. Clinic. Microbiol.* **2001**, *39*, 2531-2540.

31. Dubiley, S.; Kirillov, E.; Mirzabekov, A. Polymorphism analysis and gene detection by minisequencing on an array of gel-immobilized primers. *Nucl. Acids Res.* **1999**, *27*, E19.

32. Strizhkov, B.N.; Drobyshev, A.L.; Mikhailovich, V.M.; Mirzabekov, A.D. PCR amplification on a microarray of gel-immobilized oligonucleotides: Detection of bacterial toxin- and drug-resistant genes and their mutations. *BioTechniques.* **2000**, *29*, N4, 844-857.

33. LaForge, K.S.; Shick, V.; Spangler, R.; Proudnikov, D.; Yuferov, V.; Lysov, Yu.; Mirzabekov, A. Detection of single nucleotide polymorphisms of the human mu opioid receptor gene by hybridization or single nucleotide extension. *Am. J. Med. Genetics, Neuropsychiatric Genetics Sec.* **2000**, *96* (5), 604-615.

34. Lapa, S.; Chikova, A.; Shchelkunova, S.; Mikheev, M.; Mikhailovich, V.; Sobolev. A.; Blinov, V.; Babkin, I.; Zasedatelev, A.; Sandakhchiev L.; Mirzabekov, A. Detection and species identification of orthopoxviruses on oligonucleotide microchip. *J. Clinic. Microbiol.* **2001** (*accepted*).

35. Shick, V.; Lebed, J.; Kryukov, G. Identification of HLA DQA1 alleles by the oligonucleotide microchip method. *Mol. Biol. (Russia, English Ed.)* **1998**, *32*, 679–688.

36. Lebed, J.; Chechetkin, V.; Turygin, A.; Shick V.; Mirzabekov A. Comparison of complex DNA with generic oligonucleotide microchips. *J. Biomol. Struct. Dyn.* **2001**, *18*, 813-823.

37. Guschin, D.; Mobarry, B.; Proudnikov, D.; Stahl, D.; Rittmann, B.; Mirzabekov, A. Oligonucleotide microchips as genosensors for determinative and environmental studies in microbiology. *Appl. Environ. Microbiol.* **1997**, *63*, 2397–2402.

38. Bonch-Osmolovskaya, E.; Miroshnichenko, M.; Lebedinsky, A.; Chernykh, N.; Nazina, T.; Ivoilov, V.; Belyaev, S.; Boulygina, E.; Lysov, Yu.; Perov, A.; Mirzabekov A.; Hippe, H.; Stackenbrandt, E.; Haridon, S.; Jeanthon, C. Microbial ecology of Siberian high temperature petroleum reservoir. **2001** (*unpublished*).

39. Bavykin, S.; Mikhailovich, V.; Zakhariev, V.; Lysov, Yu.; Kelly, J.; Flax, J.; Jackman, J.; Stahl, D.A.; Mirzabekov, A.D. Biological microchip technology for discrimination of *Bacillus anthracis* and closely related microorganisms on the basis of genetic criteria. **2001** (*unpublished*).

40. Nasedkina, T.; Domer, P.H.; Zharinov, V.; Ezhkova, E.; Dormeneva, E.; Lysov, Yu.; Shick, V.; Mirzabekov, A. Identification of chromosomal translocations in hematological malignancies by hybridization with oligonucleotide microarrays. **2001** (*unpublished*).

Detection Technologies

Chapter 3

AlphaScreen™

A Highly Sensitive, Nonradioactive and Homogeneous Assay Platform for Drug Discovery High Throughput Screening, Genomics, and Life Science Research Applications in Miniaturized Format

Changjin Wang[1,4], Roger Bosse[2], Chantal Illy[2], Lucille Beaudet[2], Philippe Roby[2], Marcia Budarf[2], Kenneth Neumann[3], Juerg Duebendorfer[1], Donald Jones[3], and Daniel Chelsky[2,*]

[1]Packard BioScience Company, 800 Research Parkway, Meriden, CT 06450
[2]BioSignal Packard, Inc., 1744 rue William, Montreal, Quebec, Canada H3J 1R4
[3]Packard BioScience Company, 2200 Warrenville Road, Downers Grove, IL 60515
[4]Current address: Cellomics, Inc., 635 William Pitt Way, Pittsburgh, PA 15238

AlphaScreen™ (*A*mplified *L*uminescent *P*roximity *H*omogeneous *A*ssay) is a novel assay platform suitable for a broad range of applications in high throughput screening (HTS), genomics, and life science research and development. This proximity based, non-radioactive assay technology is highly sensitive and robust. Assays have been performed in 96/384/1536 well plates using the AlphaQuest™ Analyzer. Miniaturization does not require increasing assay component concentration. Applications include enzyme assays, protein-protein and protein-DNA interactions, a functional cAMP assay, DNA and RNA quantitation, and SNP analysis.

In order to accelerate the drug discovery process, pharmaceutical companies are now screening large compound libraries against an increasing number of therapeutic targets using highly automated systems. Enabling this increase have been the successful development of combinatorial compound libraries, the identification of new targets from functional genomic research, automated liquid handling robotics and the technological advances in assay technology (1-3). Today the pharmaceutical industry conducts hundreds of millions of individual tests every year, and the annual total could increase to several billion tests by 2005. To cope with this "explosion" of both biological targets and compounds for testing, it is necessary to miniaturize the HTS process, so that it becomes a cost effective and manageable task. Many companies are running HTS assays in 96- and 384-well plates, and a few are starting to screen in higher density and lower volume 1536-well plates or beyond (4,5).

Miniaturization strategies are also required for genomics applications, an important example being SNP (single nucleotide polymorphism) analysis. One major application of this tool will be the search for genes associated with diseases or with side effects to drug therapy. Such searches necessitate the analysis of hundreds of thousands of SNPs against many patient samples. Miniaturization and high throughput are therefore essential.

Conventional assay technologies, such as ELISA, DELFIA, and radio-isotopic filtration assays, require separation of bound fraction from the unbound by physical means, e.g. washing. These "heterogeneous" assay technologies are cumbersome, time and labor consuming, and difficult to automate and miniaturize. Such formats are especially difficult for low affinity protein-protein binding, because the binding may be too weak to withstand the dilution incurred in a washing step. Therefore, the adaptation of "mix-and-measure" or homogeneous assay technologies becomes critical for HTS and essential for miniaturization strategies. Proximity-based homogeneous assays are widely used in today's HTS applications. Scintillation Proximity Assay (SPA) and FlashPlate® microplate assays are used extensively in radiolabeled ligand-receptor binding studies (6,7). Homogeneous Time-Resolved Fluorescence (HTRF®) using lanthanide cryptate (8,9), and LANCE technology (10) using lanthanide chelates have gained popularity in recent years because they are non-radioactive, and have less assay background and interferences. Non-proximity based homogeneous assays, such as fluorescence polarization (FP) (11) and fluorometric microvolume assay technology (FMAT) (12), have also emerged into HTS applications. Not all assay technologies, however, are suitable for miniaturization, primarily due to signal intensity. Therefore, there is a great demand for a versatile homogeneous assay platform that is highly sensitive and thus can be miniaturized, is cost effective, and can be used for broad applications in drug discovery HTS, general life science research and even genomics. Here we describe a novel assay platform, AlphaScreen, which meets the above requirements.

AlphaScreen[TM] Principles

Photochemically Triggered Chemiluminescent Amplification

AlphaScreen, is a non-radioactive homogeneous assay technology based on the Luminescent Oxygen Channeling Immunoassay (LOCI) technology developed by Ullman et al (13,14). It uses two types of proprietary latex beads, a donor and an acceptor, and exploits the short diffusion distance of singlet state oxygen ($^1\Delta_g O_2$) to initiate a photochemically triggered chemiluminescent amplification reaction. The donor beads contain a photosensitizer (phthalocyanin, shown in Figure 1) that absorbs light at 680 nm and then converts ambient molecular oxygen to the excited singlet state ($^1\Delta_g O_2$). The acceptor beads contain a thioxene derivative that reacts with singlet oxygen, generating chemiluminescence upon the rapid decay of its oxidative adduct thioxetane. This chemiluminescent energy is immediately transferred to two fluorophores contained in the same beads, shifting the emission wavelength (from 370 nm) to ~600 nm (Figure 2). Because of the high concentration of the photosensitizer dissolved in the donor beads, one bead can generate up to 60,000 singlet oxygen molecules per second. This results in a great signal amplification of a molecular binding event, making AlphaScreen a highly sensitive assay. In actual biological assays, detection sensitivity of analyte at the attomol (10^{-18} mol) level has been achieved.

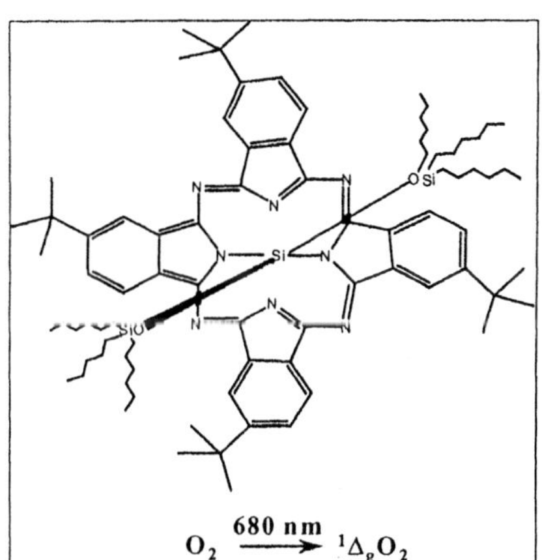

Figure 1. Structure of phthalocyanin, the photosensitizer contained in the donor beads

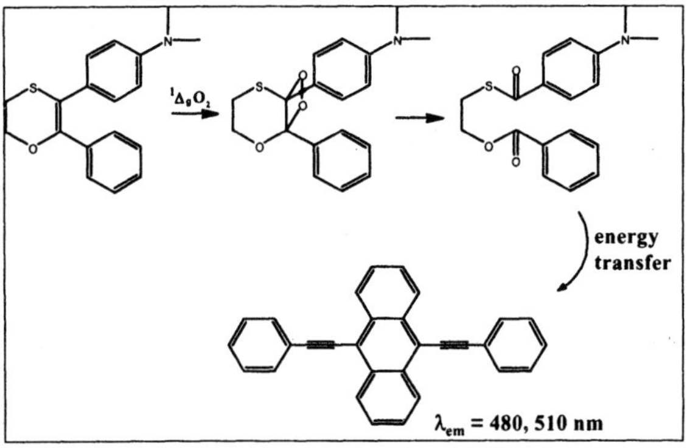

Figure 2. The cascade of the chemiluminescent reaction and the subsequent energy transfer to the one of the two fluorophores contained in the acceptor beads

Proximity-Based Homogeneous Assay

Only when the donor and acceptor beads are brought to proximity by a molecular binding event, as shown in Figure 3, will the singlet oxygen reach the acceptor beads and initiate the chemiluminescent reaction. This is due to the very short lifetime (~4 μsec.) of the singlet oxygen in aqueous solution (15), which allows diffusion to no more than ~200 nm before it decays back to ground state. On the other hand, unbound beads receive no singlet oxygen and therefore emit no light, resulting in low assay background (Figure 4). It is worth noting that this distance of 200 nm is much longer than the effective energy transfer distances of FRET and/or HTRF, making AlphaScreen compatible with large molecular complexes and indirect coupling.

Time-Resolved Fluorescence is the Detected Signal

The singlet oxygen adduct, thioxetane, decays rapidly ($k = \sim 1 \text{ s}^{-1}$) resulting in chemiluminescence energy, which is directly transferred to the fluorophores in the same beads. This makes the final AlphaScreen signal a very long lived fluorescence with a $t_1,2$ in the second range, compared to that of prompt fluorescence ($t_{1/2}$ in nsec. range) or time-resolved fluorescence ($t_{1/2}$ in μsec. range). Therefore, the AlphaScreen signal is measured in time-resolved mode with a delay between illumination and detection of 20 msec., which eliminates virtually all fluorescent background from assay components and plates. Since the illumination wavelength is very long at 680 nm, very few biological or assay substances will interfere, making AlphaScreen a highly sensitive and robust assay technology. The AlphaScreen assays are routinely run in miniaturized format at 10-25 μl volume in 384-well places and 5-10 μl in 1536-well plates. Reducing volume does not require an increase of any assay component concentrations.

Optimized Bead Size

The size of the AlphaScreen beads is optimized and uniform at 200 nm in diameter. After coating and bioconjugation, its final dimension becomes —250 nm in diameter. They are much smaller than those of other bead-based assays such as SPA beads (2-10 µm) and FMAT (6-20 µm). This feature represents significant advantages. They are too small to settle on standing in biological buffers. Thus bead suspensions can be easily dispensed using automated liquid handling devices without clogging small tips. They possess a relatively large surface area for conjugation of biomolecules and are typically used at much lower concentration (µg/ml) than that of SPA beads (mg/ml), for example. Most of the assays demonstrate near diffusion-limited kinetics, rather than solid phase reaction kinetics. Yet, they are large enough to be centrifuged and/or filtered, thus no chromatographic separation is needed for purification following bioconjugation, resulting in high yield and ease of use. The beads are very stable in suspension, even at high temperatures (e.g. 94°C for PCR), as well as in lyophilized form. They can be stored refrigerated for at least a year.

Functionalized Hydrogel Coating

The AlphaScreen beads are coated with a layer of functionalized hydrogel (a dextran derivative). This layer of hydrogel provides a hydrophilic surface to retain the dyes, minimize non-specific binding and bead self-aggregation, and provides functional groups for covalent bioconjugation. The aldehyde-activated beads are used for conjugation of antibodies, proteins, and peptides through reductive amination at very mild reaction conditions. This simple procedure is well known and results in high yield. For conjugation of nucleic acids, the iodoacetyl-activated beads and sulfhydro-derivatized nucleic acids are used at similar mild conditions. The beads have been successfully conjugated to many antibodies, proteins, peptides and nucleic acids. More than forty different assays have been developed.

AlphaScreen™ Instrumentation

To maximize AlphaScreen signal detection, the AlphaQuest™ HTS Microplate Analyzer was developed that can measure assays in 96/384/1536-well plates. The instrument uses a highly efficient laser diode emitting at 680 nm, fiber optic cables and specially optimized PMTs and can measure 10,000

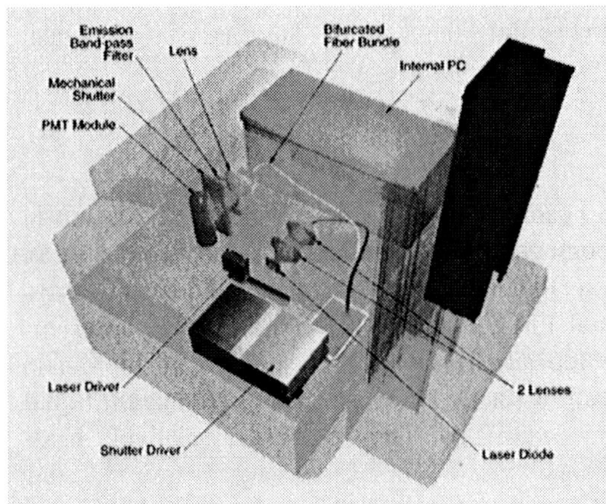

samples an hour. The optical detection assembly has also been incorporated onto the Fusion™ multifunction instrument as an assay development tool.

Figure 5. AlphaQuest™ Microplate Analyzer

Representative Applications

Many AlphaScreen assays covering a wide range of applications have been developed, including enzyme (tyrosine kinase, ser/thr kinase, proteases, helicase, etc.), cell based cAMP detection, protein-protein binding (TNFα/sTNFR1, Gβγ3y/GIRK, p53/HDM2, etc.), protein-DNA binding (estrogen receptor/estrogen response element, etc.), DNA and RNA quantitation, single nucleotide polymorphism (SNP) genotyping using allele-specific amplification or hybridization, and immunoassays. Here we describe a few representative assays. Most of the data were obtained in 384-well plate format unless otherwise stated.

Functional cAMP Assay

G-protein-coupled receptors (GPCRs) are important molecular targets for drug discovery and signal transduction studies (16). Functional responses of GPCRs may include an increase in inositol-trisphosphate (IP$_3$) turnover followed by changes in intracellular calcium concentrations or an increase or decrease in cAMP levels. As seen in Figure 6., a biotinylated cAMP derivative was designed and synthesized that is stable and capable of binding to both the streptavidin-donor bead and the anti-cAMP antibody labeled acceptor bead to form a sandwich. Increased levels of cAMP result in competition for the anti-cAMP antibody, decreasing binding of the biotinylated derivative, and thus lower counts. A representative example of cAMP quantitation following GPCR stimulation is shown in Figure 7. In this example, the basal level of cAMP in the cells is low, resulting in approximately 40,000 cps. Forskolin stimulation of the cells results in activation of adenylyl cyclase and an increase in cAMP, reducing counts to approximately 10,000 cps. Stimulation of the Formyl Peptide Receptor la (FPRIa) with the agonist fMLP (○) partially inhibited the forskolin-induced production of cAMP (●)with a Z' factor of 0.5, indicating it is suitable for HTS (17). The coefficient of variation of each data population was

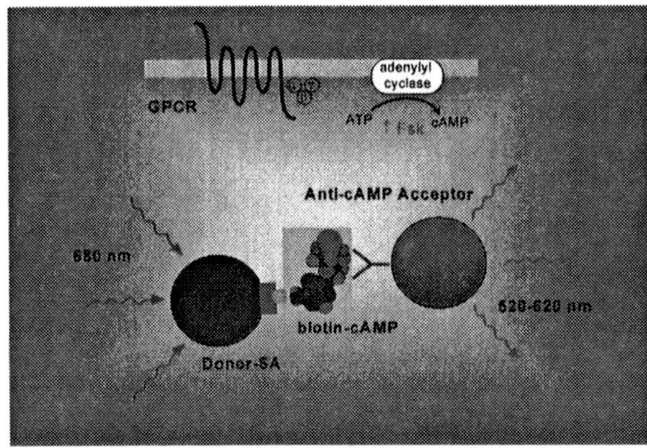

approximately 5%. The assay has also been used to measure response to receptors that result in a direct stimulation of cAMP levels, without the use of forskolin.

Figure 6. AlphaScreen™ cAMP Functional Assay

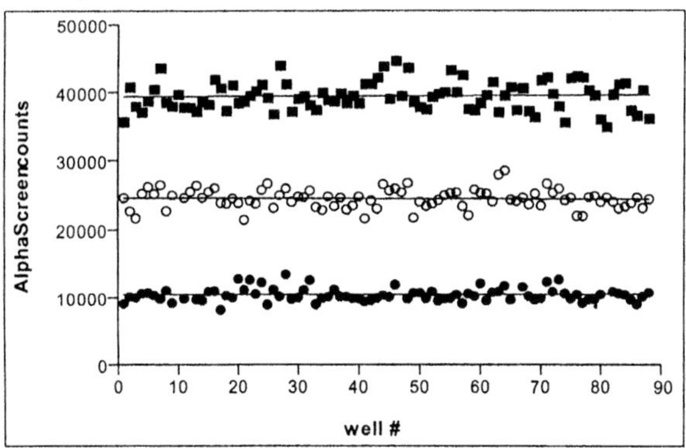

Figure 7. Three populations of data from a cAMP detection assay. Solid squares represent basal CHO cell signal. Solid circles are forskolin treated cells, and open circles are forskolin and agonist treated cells.

Nucleic Acid Detection and Single-Nucleotide Polymorphism (SNP) Genotyping

The AlphaScreen platform is an extremely sensitive and powerful technology for nucleic acid-related applications. To detect single-stranded DNA, RNA, plasmids or PCR products, donor and acceptor beads conjugated to generic oligonucleotides are used together with a pair of target-specific bridging probes (Figure 8). The bridging probes have a region complementary to a specific target and a generic tail hybridizing to the oligos conjugated to either the donor or acceptor bead. When the two probes hybridize to the same nucleic acid target, the donor and acceptor beads are brought into proximity and the AlphaScreen signal is generated. The limit of detection of the AlphaScreen technology for nucleic acids is at the low attomolar level with a linear detection range of about 300 fold. The AlphaScreen beads and probes do not interfere with enzymatic reactions and can be integrated into restriction enzyme and PCR reactions for the detection of specific targets.

The combination of AlphaScreen with PCR has allowed the development of two homogeneous SNP genotyping platforms. The first platform is based on the detection of allele-specific amplification (ASA) products. In AlphaScreen ASA, the two SNP alleles are amplified in separate wells. The 3'end of one of the PCR primers in each well is complementary to the SNP. If the allele specified by the primer is present, there is amplification of a PCR product by the Taq polymerase. If not, there is no amplification. A pair of bridging probes hybridizes to the amplified products and brings into proximity donor and acceptor beads. A signal is generated in all wells where amplification occurs.

In the second AlphaScreen platform, SNP detection is made by the allele-specific hybridization (ASH) of a bridging probe. This probe has the SNP base in the center of its target-specific portion. After PCR amplification, probe hybridization is performed at a temperature that will allow the hybridization of the matched probe but not of the mismatched probe. A common probe, which will hybridize to the amplified fragment, is also present. If both the allele-specific probe and the common probe hybridize to the same product, the donor and acceptor beads will be brought together and signal will be generated. As demonstrated in Figure 9, robust genotyping results are obtained for SNP JCNCBI DNA samples of the three genotypes (AA, AC, CC), with high signal to background ratios.

The AlphaScreen genotyping assay using ASA or ASH is a "one pot" assay combining amplification and detection in the same well. The PCR/AlphaScreen mix and genomic DNA are added into 96- or 384-well plates in final volumes ranging from 5 to 25 µl. After reagent dispensing, the plate is sealed and transferred to a thermal cycler. Following amplification and probe hybridization, the plate can be read directly on the AlphaQuest Analyzer without removing the seal. This "mix-heat-measure" operation not only eliminates the cumbersome and time-consuming sample transfer and gel electrophoresis as needed for conventional assays, but also reduces the risk of sample contamination, making it possible to run high throughput SNP genotyping assays efficiently and cost-effectively.

MAP Kinase Assay

Mitogen-activated protein (MAP) kinases are serine/threonine kinases that are activated by phosphorylation and play an important role in cellular functions. They are rapidly activated in response to various extracellular signals, such as growth factors, cytokines, and different types of cellular stress. Hence the MAP kinase family has received considerable attention in drug discovery HTS (18-19). An AlphaScreen MAP kinase assay was developed using a baculovirus expressed recombinant ERK1 enzyme (Figure 10) as a crude lysate. In this indirect format, the peptide substrate is biotinylated and captured by the

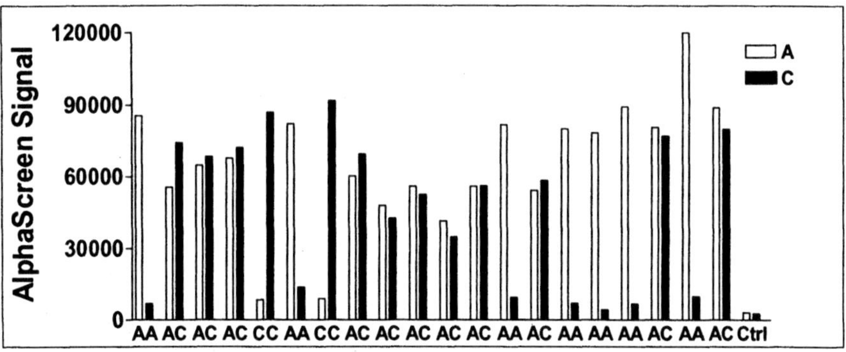

Figure 9. AlphaScreen™ ASH in 384-well format. Twenty genomic DNA samples were genotyped for the SNP marker WIAF-896 (NCBI STS Accession #G28096) in a final volume of 10 µl. After PCR amplification, the bridging probes were hybridized to their specific target at 56°C for 5 min and the beads annealed to the probe's generic tails for 20 min at 37°C. The plate was then read on the AlphaQuest™ Analyzer.

streptavidin donor bead. The phosphorylated form of the peptide substrate is then recognized by a sequence-specific anti-phosphothreonine antibody. Rather than coupled directly to the acceptor beads, the antibody is labeled with digoxigenin and captured by anti-digoxin labeled acceptor beads. Typical titration curves are shown in Figure 11.

Helicase Assay

The DNA helicases are responsible for the unwinding of double-stranded DNA and represent an important class of targets for the development of novel

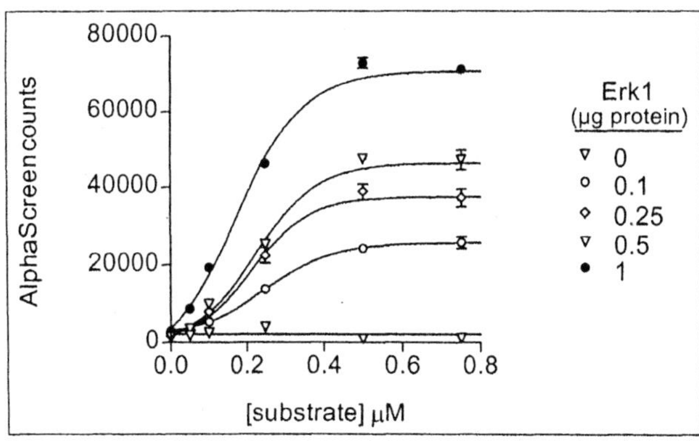

Figure 11. Titration of ERK1 and its substrate.

anti-infective agents. Conventional assay methods used radiolabeled (20) or lanthanide-labeled oligonucleotides (21). As shown in Figure 12, a sensitive AlphaScreen helicase assay was developed. The helicase activity of the SV4O large antigen is determined using a double-stranded DNA substrate, in which one of the strands is labeled with digoxigenin. Once the enzyme unwinds the substrate, a biotinylated complementary strand ("capture oligo"), present in excess, captures the digoxigenin labeled strand. Again, the streptavidin-donor and anti-digoxin acceptor beads were used to form the binding complex, yielding a convenient signal-increase assay with high sensitivity (Figure 13).

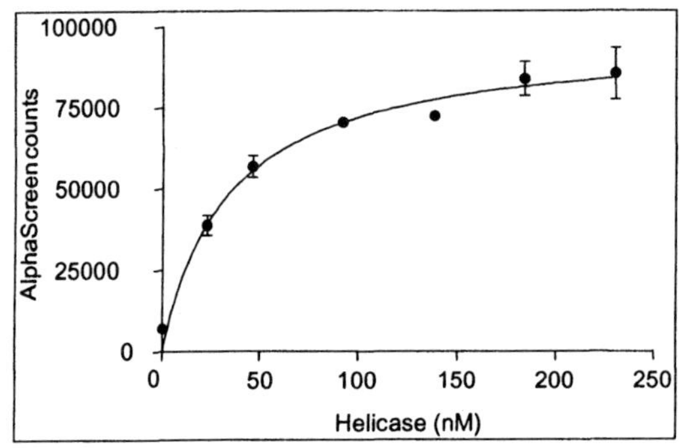

Figure 13. Helicase titration

Miniaturization in 1536-well plates

The AlphaScreen TNFα / sTNF Receptor 1 binding assay (Figure 14) was used as an example to compare the detection sensitivities in 384- and 1536-well plates. This protein-protein binding assay was constructed using a biotinylated TNFα and a non-neutralizing anti-sTNF Receptor 1 antibody. In Figure 15, the results obtained in both 384-well and 1536-well plates at various volumes are compared. In each case, the same sample was analyzed without modification. Although the signal decreases as the volume decreases, significant signal is available even at a volume as low as 3 μl. Thus, comparable results can be obtained in low microliter volume without having to increase any assay component concentration.

Summary

We developed a novel assay platform, AlphaScreen, in 96-, 384-, and 1536-well plate format. This proximity-based non-radioactive homogeneous assay platform is highly sensitive and robust, suitable for a broad range of applications

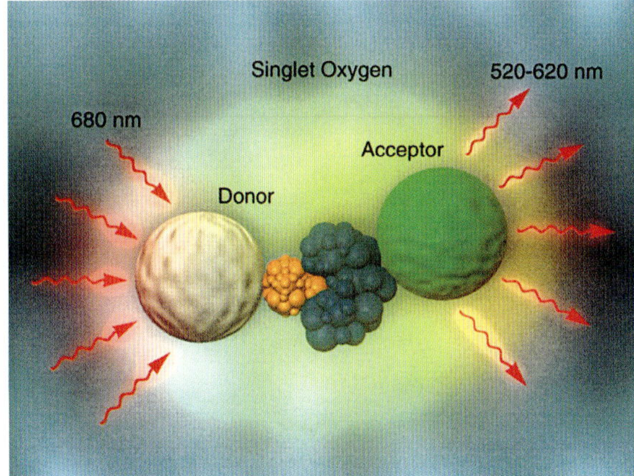

Figure 3. When AlphaScreen™ beads are brought into proximity, a highly amplified signal is generated.

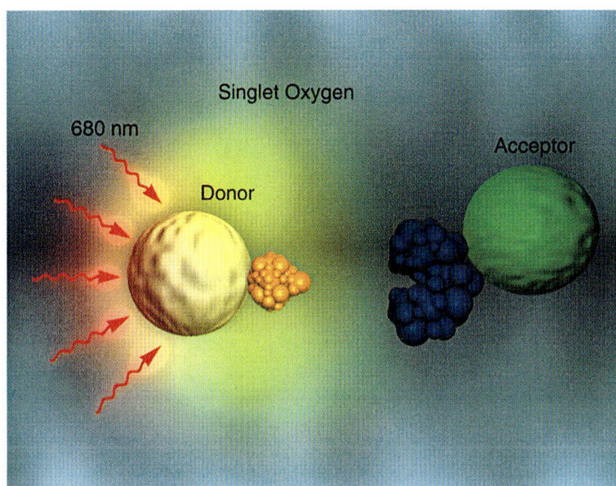

Figure 4. If AlphaScreen™ beads are not brought into proximity, no signal is generated.

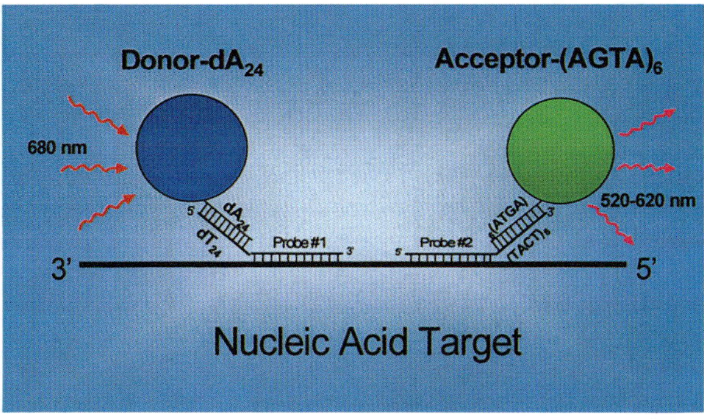

Figure 8. AlphaScreen™ detection of nucleic acids. See text for details.

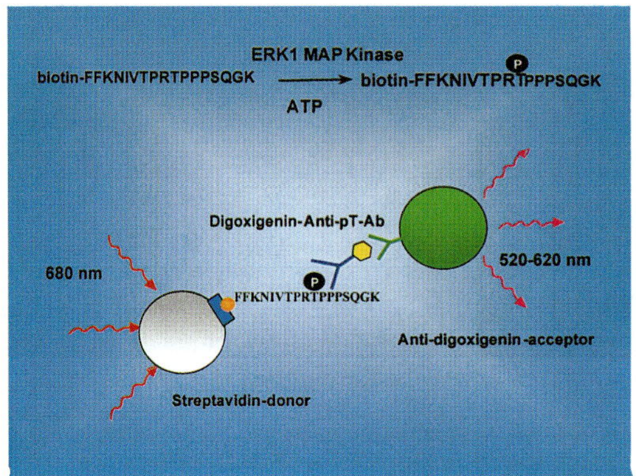

Figure 10. AlphaScreen™ MAP kinase assay using indirect capture of a phosphorylated peptid.

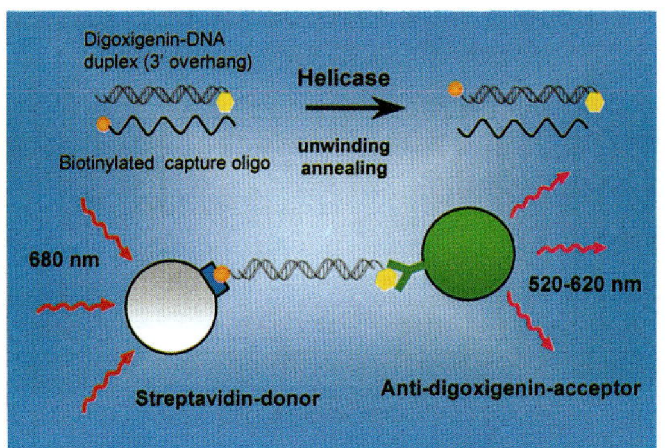

Figure 12. AlphaScreen™ Helicase Assay

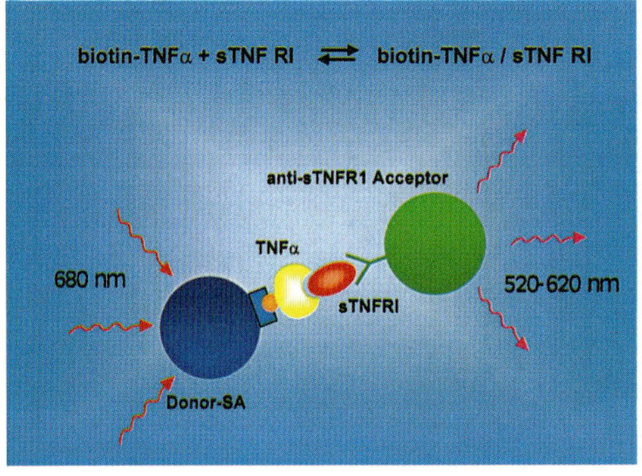

Figure 14. AlphaScreen™ TNFα/sTNFR 1 binding Assay

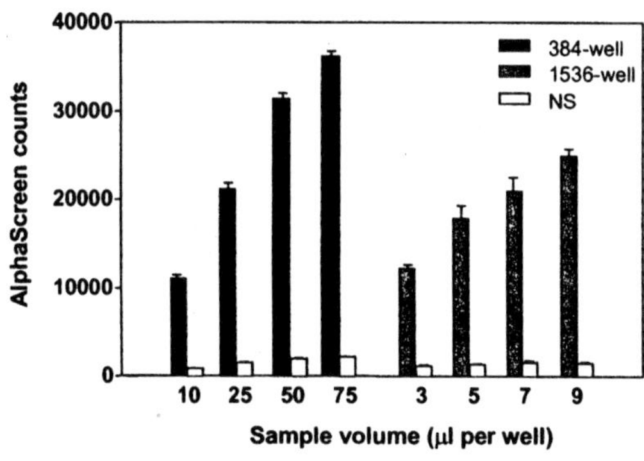

Figure 15. TNFα/sTNFR 1 binding assay in 384- and 1536-well plates

in life science research including drug discovery HTS and genomics. It can be miniaturized in 1536-well plates or beyond without having to increase any assay component concentration.

Acknowledgement

We would like to thank Drs. C. Wartchow of Targesome Co. (Palo Alto, California), S. Rose, R. Patel, and H. Kirakossian of Dade Behring, Inc. (San Jose, California), for many valuable advises and discussions.

References

1. Fox, S.; Farr-Jones, S; Yund, M. *J. Biomol. Screen.* **1999**, 4, 183-186
2. Pope, A.; Hertzberg, R. *J. Biomol. Screen.* **1999**, 4, 231-234
3. Fernandes, P. *Curr. Opin. Chem. Biol.* **1998**, 2, 597-603
4. Dunn, D.; Olowski, M.; Gastgeb, F.; Appell, K.; Ozgur, L.; Webb, M.; Burbaum, J. *J Biomol. Screen.* **2000**, 5, 177-187
5. Maffia, A.; Kariv, I.; Oldenberg, K. *J. Biomol. Screen.* **1999**, 4, 137-141
6. Graziani, F.; Aldegheri, L.; Terstappen, G. *J. Biomol. Screen.* **1999**, 4, 3-7
7. Bosse, R.; Garlick, R.; Brown, B.; Menard, L. *J. Biomol. Screen,* **1998**, 3, 285-292
8. Zhou, G.; Cummings, R.; Li, Y.; Mitra, S.; Wilkinson, H.; Elbrecht, A.; Hermes, J. *Mol. Endocrinol.* **1998**, 12, 1594-1604
9. Mellor, G.; Burden, M.; Preaudat, M.; Joseph, Y.; Cooksley, S.; Ellis, J.; Banks, M. *J. Biomol. Screen.* **1998**, 3, 91-99
10. Hemmila, I. *J. Biomol. Screen.* **1999**, 4, 303-307
11. Banks, P.; Gosselin, M.; Prystay, L. *J. Biomol. Screen.* **2000**, 5, 159-167

12. Martens, C.; Bakker, A.; Rodriguez, A.; Mortensen, R.; Barrett, R. *Anal. Biochem.* **1999**, 273, 20-31

13. Ullman, E.; Kirakossian, H.; Singh, S.; Wu, P.; Irvin, B.; Pease, J.; Switchenko, A; Irvine, J.; Dafforn, A.; Skold, C; Wagner, D. *Proc. Natl. Acad. Sci. USA.* **1994**, 91, 5426-5430.

14. Ullman, E.; Kirakossian, H.; Switchenko, A.; Ishkanian, J.; Ericson, M.; Wartchow, C.; Pirio, M.; Pease, J.; Irvin, B.; Singh, S.; Singh, R.; Patel, R.; Dafforn, A.; Davalian, D.; Skold, C.; Kurn, N.; Wagner, D. *Clin. Chem.* **1996**, 42/9, 1518-1526

15. Rodgers, M.; Snowden, P. *J. Am. Chem. Soc.* **1982**, 104, 5541-5543

16. Kariv, I.; Stevens, M.; Behrens, D.; Oldenberg, K. *J. Biomol. Screen.* **1999**, 4, 27-32

17. Zhang, J.; Chung, T.; Oldenberg, K. *J. Biomol. Screen.* **1999**, 4, 67-73

18. Pelech, S.; Sanghera, S. *Trends Biochem. Sci.* **1992**, 17, 233-238

19. De Witt, R.; Boonstra, J.; Verkleij, A.; Post, J. *J. Biomol. Screen.* **1998**, 3, 277-284

20. Kyono, K.; Miyashiro, M.; Taguchi, I. *Anal. Biochem.* **1998**, 257, 120-126

21. Earnshaw, D.; Moore, K.; Greenwood, C.; Djaballah, H.; Jurewicz, A.; Murry, K.; Pope, A. *J. Biomol. Screen.* **1999**, 4, 239-248

Chapter 4

Light-Powered Microtransponders for High Multiplex-Level Analyses of Nucleic Acids

Wlodek Mandecki, Michael G. Pappas, Natan Kogan, Zhuying Wang, and Beata Zamlynny

PharmaSeq, Inc., 11 Deer Park Drive, Suite 204, Monmouth Junction, NJ 08852–1923

We have developed light-powered microtransponders, i.e. miniature electronic transmitters, composed of photovoltaic cells, read-only memory (ROM), logic, antenna and circuitry necessary to transmit unique radio frequency (RF) signals in real time, for use in rapid nucleic acid analyses. Microtransponders are extremely small integrated circuits, 500μ on each side and even smaller units are being developed (250μ). Nucleic acid probes covalently linked to microtransponder surfaces hybridize to fluorescence-tagged, complementary strands of nucleic acid. When assayed in a flow fluorometer, each microtransponder will transmit a unique identification number to a nearby receiver, identifying the exact nucleic acid probe sequence. When target DNA hybridizes to the probe on the microtransponder surface, both an the identification signal and a fluorescence signal are generated when the microtransponder is struck by laser light. Microtransponders are prepared using a standard CMOS process (0.35μ feature size), have 50 bits of memory, and transmit three identification signals per read with a bit rate of 200 kbits per second. Light-powered microtransponders offer a novel, specific and rapid method for multiplex analysis of nucleic acids and have applications in DNA probe assays, single nucleotide polymorphism detection, drug discovery, and immunoassays.

Introduction

PharmaSeq, Inc. is developing a unique, high-throughput technology for tagging and accurately reporting the presence of DNA, RNA, antibodies, or low molecular weight drug molecules by RF (*1-3*). This novel platform technology uses tiny silicon microtransponders, each complete with circuitry allowing it to act as a radio-transmitter when activated by light. Nucleic acid probes are covalently linked to the surface of each microtransponder in such a way that each microtransponder has a large number of a specific probe molecule. When these coated microtransponders are incubated in test specimens containing tagged (fluorescent) target nucleic acid molecules, complementary target molecules specifically hybridize to the probes on the microtransponder. As little as one nucleotide mismatch will prevent hybridization of the target DNA to the probe. After washing the microtransponders to remove all non-specifically bound nucleic acid strands, target molecules can be detected on each microchip by observing fluorescence emitted by dyes previously attached to the target molecules. Thus, the laser light not only activates every microtransponder to emit its unique identification number but also causes a subset of microtransponders hybridized with target molecules to fluoresce brightly (positive result). The multiplex system described here is composed of specially coated microtransponders, high-throughput reading devices and system software.

Tagging Systems

Currently, many tagging systems are being used to track or identify molecules important in research and medicine (Table I). These include genes, peptides, proteins, nucleic acid strands and small organic molecules. Tags include dyes or combinations of dyes, electronic tags, electrophoric tags, mass tags, nucleotides, and peptides; also "x, y" (Cartesian) coordinates may be used to identify molecules on a planar surface (*4,5*). Each tag must have a way of being decoded. Therefore, assays or instruments are used to determine the presence or absence and identity of each tag. However, many tags require tedious or lengthy assay procedures, expensive instruments, or both in order to track or identify target molecules reliably. Assays using some tagging systems take many hours to perform, increasing the labor cost per test and severely limiting the numbers of assays that can be performed in the laboratory.

Encoding Methods in DNA Assays

Two broad methods of encoding DNA are used in nucleic acid-based assays. They can be called, in simple terms, two-dimensional (2-D) and three-

Table I. Tags Are Commonly Used to Track Many Types of Molecules, such as Nucleic Acids, Peptides, Proteins or Small Organic Molecules

Tags
Combination of fluorescent dyes
Electrophoric tag
Mass tag
Nucleotide
Peptide
Secondary amine tag
X, Y coordinates

dimensional (3-D) -type encoding. The methods differ in whether the DNA probes are present in the assay (typically during hybridization) in a 2-D planar arrangement, or as a 3-D configuration in space (i.e. volume of a test tube). 2-D DNA encoding is accomplished by attaching DNA molecules in arrays at specific x and y coordinates onto a flat surface. A popular attachment surface for 2-D nucleic acid arrays is a silane-treated glass microscope slide. Hundreds to thousands of unique molecules can be spotted on the treated glass surface in very small (picoliter to microliter) volumes and each identified by its Cartesian (x, y) coordinates. Over one hundred companies are involved worldwide in some way to prepare or detect molecules using 2-D arrays (reviewed in *6*).

Encoding the DNA probes in a 3-D configuration presents a greater challenge. Two basic approaches are being pursued in DNA assays, namely (a) the use of fluorescent dye-impregnated microparticles, or (b) the use of the PharmaSeq's light-powered microtransponders (*1-3*). The use of dye-encoded microbeads is championed by Luminex, Inc. (*7*). In this approach, the beads are encoded with a few fluorescent dyes at a several concentrations. For instance, if two different dyes having different excitation/emission properties are used at ten different concentrations, then the number of possible codes is 100 (10^2) combinations. The limitation of the approach is the number and concentrations of fluorescent dyes that can be used in any one assay. Too many variations in dyes or their intensities can limit meaningful statistical analysis of the results and prevent determination of positive or negative results.

There are several advantages of the 3-D over a 2-D assay approach. The 3-D assay tends to be more compact. If a traditional microarray assay employing, let's say, 10,000 probes, requires an area that is 1-10 cm across, an equivalent microbead assay using six micron diameter beads would require a volume approximately 200 microns across, or with the microtransponders having dimensions 250 x 250 x 100 microns, roughly 0.5 cm across. Typically, there is more flexibility in choosing the solid support in the 3-D-type assay, the process of depositing or synthesizing the probe on a bead or microtransponder can be controlled better, and mixing during hybridization steps can be more effective. Probe immobilization methods on microtransponders are analogous to those employed in bead-based immunoassays, and therefore, cost-effective. In addition, microtransponders are very well suited for combinatorial chemistry methods used to synthesize DNA strands. Therefore, interest in 3-D-type DNA-based assays has been growing.

Microtransponders

Microtransponders developed at PharmaSeq are tiny, monolithic, light-powered devices. Each is a fully integrated circuit containing photovoltaic cells, synchronization logic, ROM containing identification data, read logic, a modulator, and a loop antenna (Figure 1). Microtransponders are manufactured using Very Large Scale Integration (VLSI) silicon-based technology, which is also used for manufacturing computer processors and memory chips. All the above micro components are fabricated on the surface of silicon wafers using CMOS processing (0.35 micron feature size).

The wafers are then diced using lasers to produce tens of thousands of tiny silicon microtransponder cubes. Microtransponder cubes on the order of 500 microns have been manufactured and found to be fully functional. A single eight inch diameter silicon wafer can yield several hundred thousand microtransponders. Microtransponders having the size of 250 microns are being developed. Each microtransponder can store information identifying the nucleotide sequence of a nucleic acid probe in its ROM.

Receiver

The RF signal generated by the microtransponders are picked up by the receiving unit. The receiver is a multi-component system composed of an RF pickup magnetic assembly, a low noise amplifier, a band pass filter, a real time analog-to-digital signal processor, and a microprocessor (Figure 2). The

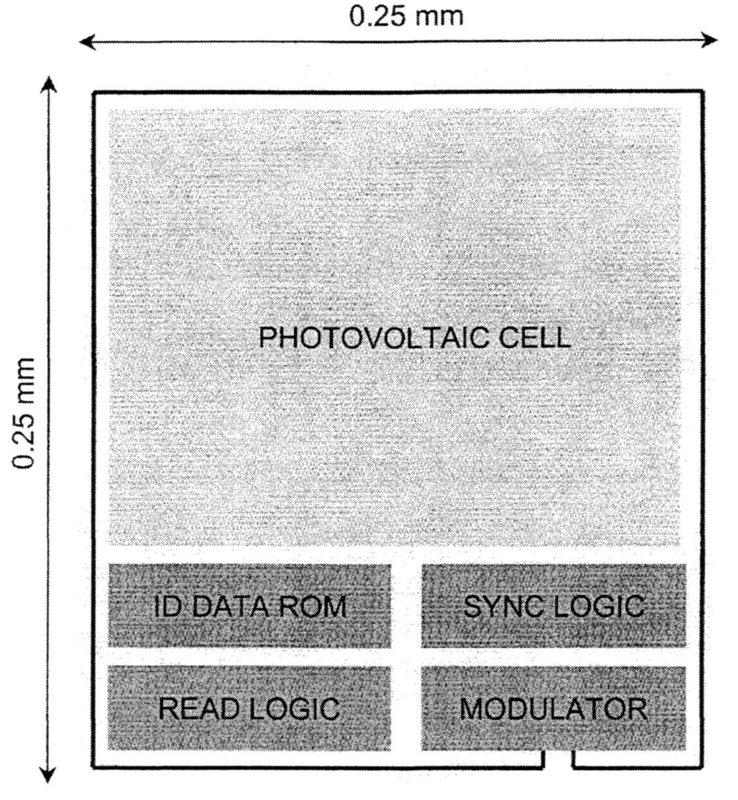

Figure 1. Microtransponder block diagram.

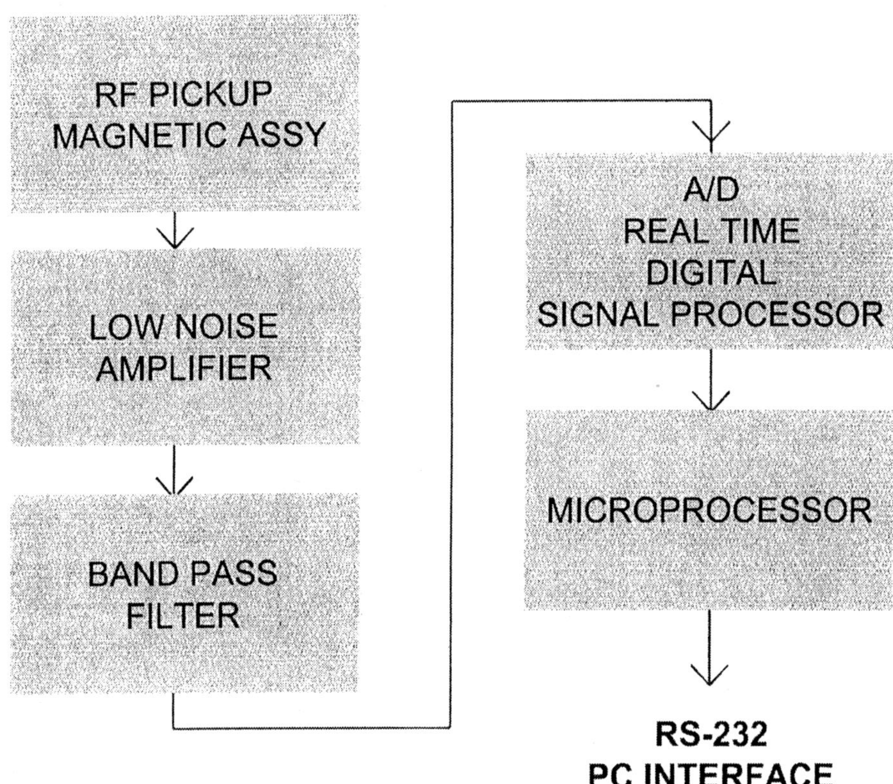

Figure 2. Receiver block diagram.

microprocessor receives microtransponder RF transmissions as bits of data (0s or 1s) and then transmits this data to a computer system with appropriate software to analyze, tabulate, store and print the information. A typical microtransponder signal generated by a light-activated microtransponder is shown in Figure 3. Transmission to a computer for bit processing and storage can be accomplished using a standard RS-232 computer interface, the newer universal serial bus (USB) port or by wireless devices.

Presently, the distance between a microtransponder and the RF receiving unit is critical. The RF antenna must be approximately 2-4 mm from the microtransponder in order to receive adequate signal. This distance can be lengthened by increasing light intensity, modifying the antenna configuration, or by changing the electronic circuitry. Antenna characteristics are shown in Figure 4. The receiver antenna can be a simple ferrite core wrapped in copper wire. Another factor in signal generation is the angle at which microtransponders are struck by light. Although laser light striking the microtransponder perpendicularly gives the highest energy transfer, laser light can strike the microtransponder at an angle of up to $\pm 60°$ and still generate enough power to produce a readable RF signal. In addition, redundancy is built into the identification system. Up to three reads can be performed in one millisecond, assuring that each microtransponder is correctly identified by the receiver. These can be an important factors in receivers used for high throughput flow systems. In addition, under these experimental conditions the system has an excellent signal to noise (S/N) ratio of ≥ 20 dB.

Flow Analysis System

In the microtransponder system presently being developed, we envision analyzing test samples under continuous high-throughput flow conditions (flow fluorometry). The microchips would flow in a single file through a narrow diameter capillary tube, past a laser beam. Laser light would power each microchip, causing it to transmit its radio frequency identification number to a pick-up antenna wrapped around the capillary tube. The same laser light would also cause all fluorescent dye-labeled target DNA sequences attached to the microchip to fluoresce. Fluorescence would be read by a photomultiplier tube in the flow fluorometer. Thus, every specific probe sequence will be identified along with a positive or negative fluorescence result. In practice, this will be done by matching the fluorescence signals with the identification number of the microchip. All data from this high-throughput assay will be stored in a computer database and analyzed. Signal-to-background ratios in hybridization experiments using microtransponder emulators have been very high ($\geq 100:1$). High S/N ratios help to distinguish between truly positive and negative DNA

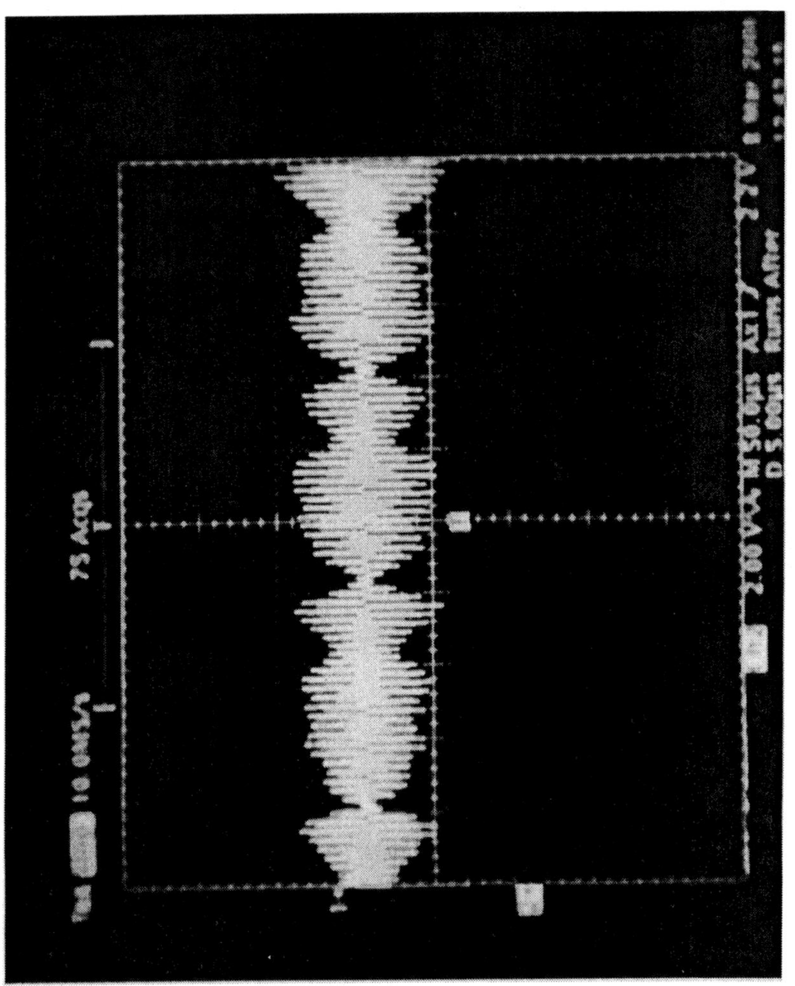

Figure 3. Typical microtransponder signal.

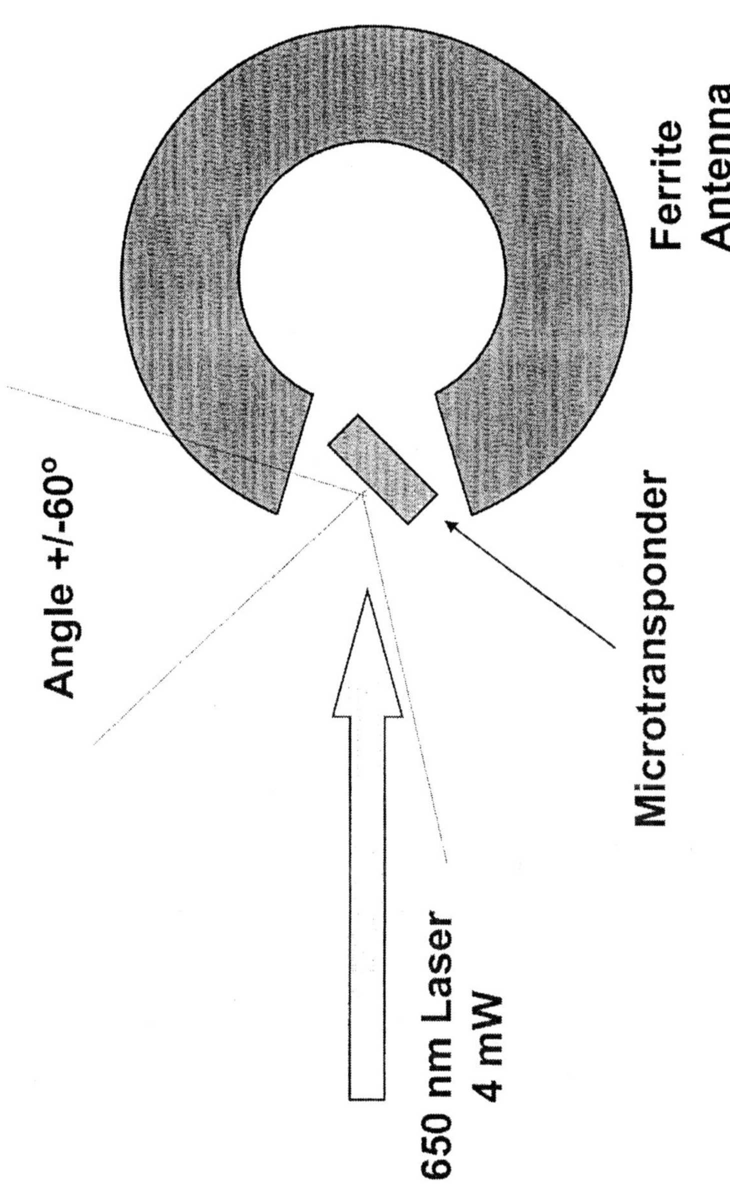

Figure 4. Receiver antenna characteristics.

test samples. This system can potentially detect, differentiate and report up to thousands of probes and hybridized target molecules per second in a given user-defined assay, similar to the rates achieved in flow cytometry.

Gene Analysis Using DNA Hybridization

We developed a hybridization assay to detect mutations in the protease (*pr*) gene of a strain of human immunodeficiency virus type 1 (HIV-1, NY5/IIIB, recombinant clone PNL4-3)(*8*). This assay was developed because single nucleotide mutations in the *pr* gene sequence can confer resistance to currently prescribed anti-protease drugs (*9*). The purpose of the experiments was to determine whether silicon microchips with surface characteristics similar to microtransponders and coated with DNA probes could successfully bind to complementary and mismatched target DNA under standard hybridization conditions.

Emulators, silicon microchips with the same surface characteristics as microtransponders, were manufactured and cut into 500 micron cubes. Emulators are routinely produced without logic circuitry, but are only able to transmit a radio frequency when activated by light. Microchip emulators were washed and silane-treated identically to fully functional microtransponders. Biochemical linkers were attached to microchip emulator surfaces in order to allow covalent coupling of primary amine-labeled oligonucleotide probes. Emulators were divided equally and placed in three polypropylene tubes. Three groups of oligonucleotide probes were prepared and coupled to emulators. One group contained a nucleotide sequence from the *pr* gene of HIV-1 perfectly complementary to a fluorescent dye-labeled *pr* target DNA sequence. The second group contained the nucleotide sequence from the *pr* gene with a single nucleotide mismatch. The third group contained an oligonucleotide probe composed of a random nucleotide sequence.

Microchip emulators coated with oligonucleotide probes were placed in three 1.5 mL conical polypropylene tubes (Eppendorf) and incubated in prehybridization solution at 55°C for 10 minutes. Prehybridization buffer was removed from the microchips by pipeting. Fluorescent dye-labeled target *pr* DNA solution was then added to all the tubes containing the three types of microchip emulators. Emulators were placed in a hybridization oven and rotated for 20 minutes at 55°C. All three groups of microchip emulators were then washed three times in wash buffer preheated to 55°C in order to remove nonspecifically bound DNA. The biomedical methods are essential, and are described (*10*). After washing, microchip emulators were dried and arranged in groups of three on fluorescence-compatible glass slides (Fluoroslide, Esco). For standard Cy3-labeled oligonucleotide solutions, a Fluoroware black plastic holder with square wells was used (dimensions ~500 x 500 x 150 microns). Epifluorescence emitted from the surfaces of the hybridized emulators, having an emission maximum ~570 nm and a fluorescent excitation maximum ~550 nm was observed using a Nikon Eclipse E600 fluorescence microscope equipped with a Super High Pressure Mercury Lamp, Plan Fluor objective lenses, a narrow-band FITC filter block. a Nikon C-LP Halogen lamp and a mirror block for observations under reflected white light. Microtransponder images were captured using a SenSys air-cooled, charge-coupled device (CCD) camera (Photometrics) with IPLab Scientific Imaging Software for Windows (Scanalytics) installed in a Dell desktop computer.

Data from three experiments performed on separate days are shown in the Table II. When fluorescent dye-tagged target DNA was incubated with microchip emulators having no mismatches (complementary DNA), a high level of hybridization occurred (2,447, 2,213, 2,040 pixel values). In contrast, incubation of tagged *pr* gene sequences to microchip emulators with probes having only one base pair mismatch (single mismatch DNA) resulted in almost no hybridization. The average pixel values (138, 134, 136) were just slightly higher than signals generated when a random sequence of DNA (negative control) was incubated with target HIV *pr* DNA (116, 113, 112).

Table II. Tagged HIV *pr* Gene Sequences Hybridize to Microtransponder Probes Depending Upon Sequence Complementarity.[a]

Experiment Number	*Fluorescence from Microchip Emulators Coated With:[b]*		
	Complementary DNA Probe	*Single Mismatch DNA Probe*	*Random Sequence DNA Probe*
1	2,447	138	116
2	2,213	134	113
3	2,040	136	112

[a] Exposure time = 0.5 seconds, fluorescence levels shown as pixel values
[b] Average pixel value of three microchip emulators per group in each experiment

Summary

Present research at PharmaSeq focuses on DNA probe diagnostics using microtransponders.   Light-powered microtransponders treated to allow attachment of DNA probes are an excellent solid phase for DNA hybridization. To this end, experiments showed that microtransponder emulators hybridized with exact nucleic acid segments to the HIV-1 protease *(pr)* gene fluoresce brilliantly compared to single nucleotide base pair mismatch controls and random sequence probes which failed to fluoresce (no hybridization). Single nucleotide differences in viral genes is important in drug discovery and in patient monitoring during treatment. In addition, due to their very small size, relatively large memory size (50 bits) and fast data transmission potential, they can be readily used in a high-throughput flow system to assay thousands target molecules injust a few minutes. Quite importantly, PharmaSeq's 3-D approach allows the choice of any microchip(s) to create any cocktail of interest. This "microchip cocktail" leads to a precise analysis of target molecules. Assays are user-defined and provide unlimited applications to the investigator, and offer numerous advantages over standard 2-D nucleic acid probe assays The ability to tag and monitor individual molecules or objects using light-powered microtransponders offers unprecedented utility in a variety of areas such as immunoassays, combinatorial chemistry, genomics, and single nucleotide polymorphism (SNP) detection.

References

1.   Mandecki, W. U.S. Patent 5,641,634, 1997.

2.   Mandecki, W. U.S. Patent 6,001,571, 1999.

3.   Mandecki, W. U.S. Patent 6,046,003, 1999.

4.   Gallop, M. et al. Applications of combinatorial technologies to drug discovery. 1. Background and peptide combinatorial libraries. *J. Med. Chem.* **1994**, *37*, 1233-1249

5. Meza, M.B. Bead-based HTS applications in drug discovery. *Drug Discovery Today* **2000**, *1*, 38-41.

6.   Abramowitz, S. DNA analysis in microfabricated formats. *J. Biomed. Microdevices* **1999**, *1*, 107-112.

7. Fulton R.J., McDade R.L., Smith P.L., Kienker L.J., Kettman J.R., Jr. Advanced multiplexed analysis with the FlowMetrix system. *Clin. Chem.* **1997,** *43,* 1749-56.

8. Shafer, R.W.; *et al.* Sequence and drug susceptibility of subtype C protease from human immunodeficiency virus type 1 seroconverters in Zimbabwe. *AIDS Resch. Human Retrovir.* **1999,** *15,* 65-69.

9. Adachi, A.; *et al.* Production of acquired immunodeficiency syndrome associated retrovirus in human and nonhuman cells transfected with an infectious molecular clone. *J. Virol.* **1986,** *59,* 284-291.

10. Guo, Z., Guilfoye, R.A., Thiel, A.J., Wang, R., Smith, L.M. Direct fluorescence analysis of genetic polymorphisms by hybridization with oligonucleotide arrays on glass supports. *Nucleic Acids Research* **1994,** *22,* 5456-5465.

Acknowledgements

We wish to acknowledge the outstanding efforts of the engineering teams at Sarnoff Corp. and Tachyon Semiconductor Corp. for design and production of microtransponders and receivers. This research was funded by the Advanced Technology Program of the National Institute of Standards and Technology, The National Human Genome Research Institute, and the State of New Jersey Commission on Science and Technology.

Chapter 5

DNA Biosensor: Immunosensor Applications for Anti-DNA Antibody

Koji Nakano[1,2], Takahiro Anshita[1], Masamichi Nakayama[1], Hiroshi Irie[1], Yoshiki Katayama[1], and Mizuo Maeda[1]

[1]Department of Applied Chemistry, Faculty of Engineering, Kyushu University, 6–10–1, Hakozaki, Fukuoka 812–8581, Japan
[2]PRESTO, Japan Science and Technology Corporation (JST), 4–1–8 Honchou, Kawaguchi 332–0012, Japan

Two examples of DNA biosensor that respond selectively to anti-DNA antibody are reported. These biosensors comprise a double-stranded DNA as a receptive component and a redox-active, indicator molecule for electrochemical detection. First, a chemical modification with 2-hydroxyethyl disulfide was made on the terminal phosphates of calf thymus DNA and the DNA was immobilized onto Au surface *via* chemisorption. By taking advantage of the principle of an "ion-channel sensor", the DNA-modified electrode was successfully applied for the detection of anti-DNA antibody (mouse, monoclonal IgM); cyclic voltammetric measurements for $[Fe(CN)_6]^{4-/3-}$ with the DNA-modified electrode gave reversible current responses in the range of 1 — 100 nM of the IgM while neither mouse IgM nor mouse IgG did not give such responses. Next, the chemoreceptive surface adlayer was improved to a DNA/ferrocene bilayer format to give an analyte-dependent electrochemical signal. The sensor also responded to the anti-DNA antibody in the range of 1 — 10 nM without the use of additional indicator molecules. These DNA biosensors would be useful for monitoring DNA—protein interactions if the immobilized DNA has a specific sequence to the target protein.

The Human Genome Project has continued to achieve substantial progress in sequencing chromosomal DNA. With this enormous body of information, intensive efforts have already been launched in relation to genomics—the study of the structure and function of genetic material in chromosomes (*1*). From the viewpoint of chemical sensor development, it should be pointed out that sequence-specific methods for detecting nucleic acids are critical for such purposes. Incidentally, interaction between proteins and nucleic acids occur at all levels of DNA replication, expression and in numerous regulatory processes. These regulated events culminate in the appearance of a functional gene product in the cell, and are therefore of the greatest importance in life. Proteomics, which involves the systematic analysis of the protein expression, has become an important area spanning many diverse fields (*1*).

The interactions between proteins and DNA are characterized most notably by their stability and exquisite specificity; equilibrium dissociation constants in the order of 10^{-9} — 10^{-12} M and ratios of specific to nonspecific binding of 10^3 — 10^7 are typical (*2*). With this understanding, one may notice the possibility of developing a biosensor that comprises a DNA as a receptive component. The sensor would provide a certain means for the evaluation of DNA—protein interactions. The initial hurdle to overcome is the stable attachment of DNA on a transducing element which commonly includes a metallic electrode. Next, it can be very difficult to observe and characterize noncovalent binding between DNA and proteins since neither of them have unique spectroscopic characteristics nor a oxidative/reductive capacity in most cases. One approach used to observe the DNA—protein binding involves the introduction of a functional group into the DNA that can report on complex formation (2).

We have been investigating chemically-modified electrode applications of DNA; terminal phosphates were chemically modified with an organosulfur precursor and the DNA double-strands were immobilized on a Au electrode *via* chemisorption (Scheme 1). By taking advantage of the principle of an "ion-channel sensor" (*3*), the DNA-modified electrode was successfully used for detecting DNA-binding molecules and ions (*4-6*). In the present paper, a possible application of the DNA biosensor for monitoring DNA—protein interactions was studied using an anti-DNA antibody as the exploratory model system. The sensor responded selectively to the antibody in the concentration range of 1 — 100 nM. Next, the chemoreceptive surface adlayer was improved to a DNA/ferrocene bilayer format (Scheme 2) to give an analyte-dependent electrochemical signal. The sensor was found to respond to the antibody without the use of a reporter molecule.

Experimental

Chemicals and Biochemicals

The following reagents were commercially available, of the highest grade and used as received: 1,1'-Ferrocenedicarboxylic (Fc) acid, 2-hydroxyethyl

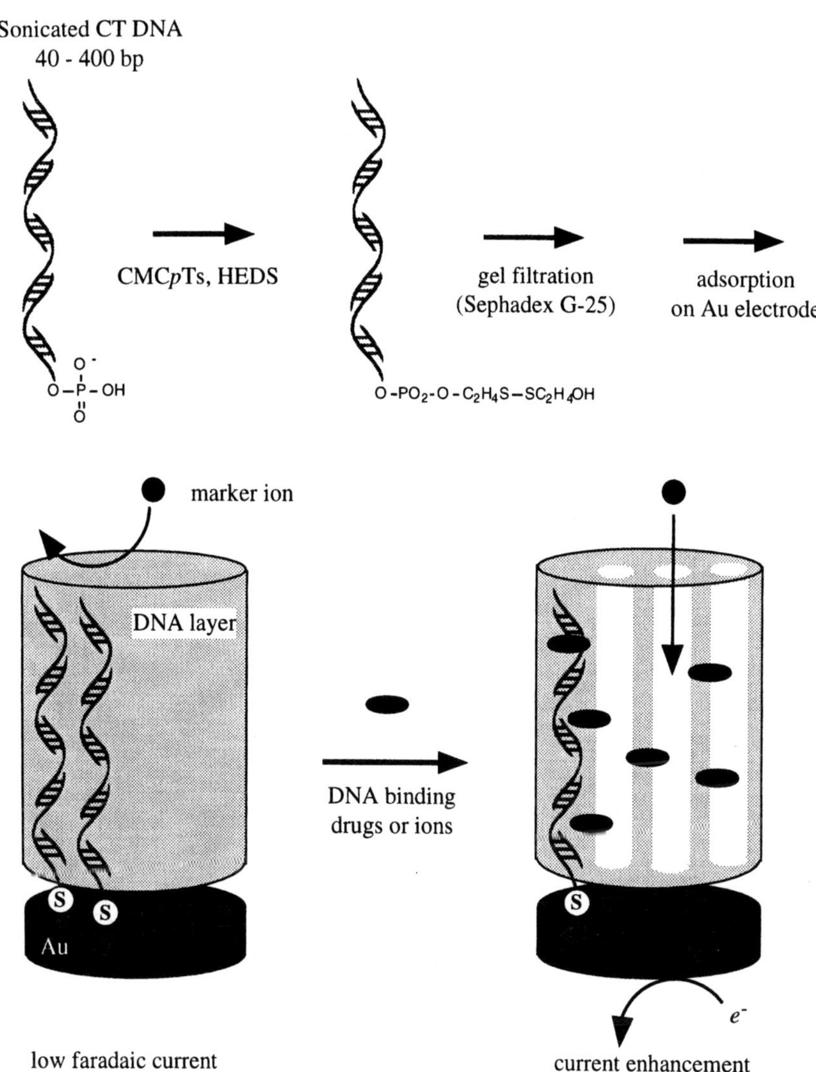

Scheme 1. Electrode preparation and DNA-dependent ion-channel mechanism.

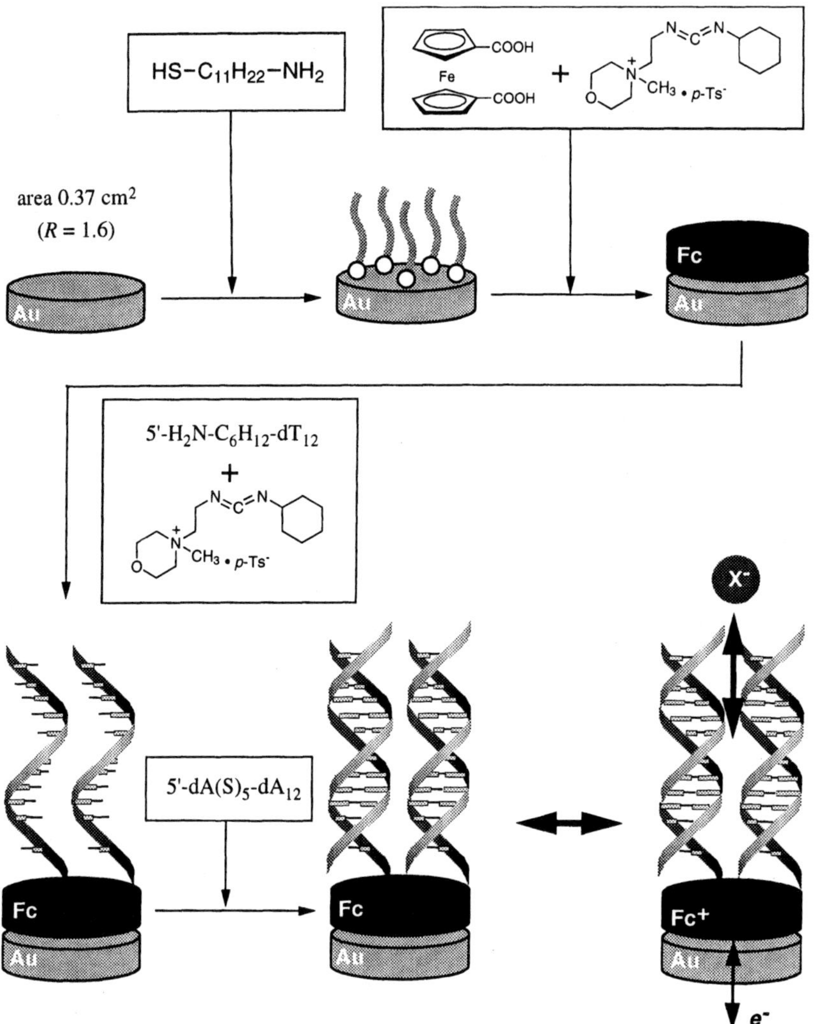

Scheme 2. Stepwise derivatization of Au electrode surface for the preparation of DNA-modified electrodes with DNA/Fc/Au interfacial structure.

disulfide (HEDS), 2-mercaptoethnol (ME) and 1-cyclohexyl-3-(2-morpholino-ethyl)carbodiimide metho-*p*-toluene-sulfonate (CMC*p*Ts). 11-Aminoundecane-thiol (AUDT) hydrochloride was prepared from 11-bromoundecanoic acid *via* a crystalline isothiuronium salt as described elsewhere (*7*). The calf-thymus (CT) deoxyribonucleic acid sodium salt (Type I) was obtained from the Sigma Chemical Co. The CT-DNA was first subjected to sonication and the DNA fragments were purified by the standard procedure. Agarose gel electrophoresis showed that the molecular weight of the DNA fragment was 30 — 260 kDa which was equivalent in base pairs (bp) to 40 — 400 (1 bp = 660 Da). Synthetic oligodeoxyribonucleic acids (ODNs) were obtained from commercial synthetic services. Immunoproteins, monoclonal anti-DNA IgM (mouse), mouse IgM (normal plasma) and mouse IgG (normal plasma) were purchased from Cosmo Bio Co., Ltd., and used as received. All other chemicals were guaranteed reagents and were used without further purification.

Preparation of DNA-Modified Electrode with DNA/Au Structure

Sonicated CT DNA (10 mg) was reacted with HEDS (4.6 mg) in the presence of CMC*p*Ts (0.4 g) in 0.4 cm^3 of MES buffer solution (0.04 M 2-(N-morpholino)ethanesulfonic acid–NaOH, pH 6.0, 1 M = 1 mol dm^{-3}) for 24 h at 25 °C. The reaction mixture was gel-filtered on an NAP-10 column (Pharmacia), and the macromolecular fractions, which displayed the characteristic absorption at 260 nm due to nucleic bases, were collected. The combined fractions containing the modified DNA (0.3 mM in bp concentration) was stored at 5 °C and used directly for the following procedures (*4-6*).

A gold disk electrode (1.6 mm diameter, Bioanalytical Systems Co.) was polished to a mirror finish with alumina powders of successively finer grades (Buehler Co.), and used as the working electrode. The electrode was first immersed into the disulfide-modified DNA solution for 24 h at 5 °C, and was further treated with aqueous 1 mM ME solution for 24 h at 5 °C to mask the uncovered, bare Au surface. The electrode was then washed with and stored in TE buffer (10 mM Tris-HCl, 1 mM EDTA; pH 7.2) at 5 °C before and between uses. Cyclic voltammetric (CV) measurements were made using a BAS Co. Model CV50W potentiostat. A conventionally designed three-electrode cell was used with a Pt plate (10 × 10 mm) counter electrode and a standard Ag/AgCl (3 M NaCl) reference electrode. Unless otherwise indicated, measurements were made at 25 mV s^{-1} on 10 mM each of ferrocyanide/ferricyanide solutions containing 0.1 M KCl. All experiments were made at a constant temperature of 25 °C.

Preparation of DNA-Modified Electrode with DNA/Fc/Au Structure

Self-assembled monolayers (SAMs) from an ω-aminoalkanethiol were used as the base adlayer and a stepwise derivatization was made using a coupling reagent to build the interfacial structure. Polycrystalline Au films (2000 ± 200 Å) prepared by evaporation onto glass plates were used as substrates. A gold substrate was placed in a one-chamber cell containing a horizontally mounted electrode (0.37 cm^2 exposed electrode area). The cell was filled with an aqueous 0.1 M H_2SO_4 solution (0.01 M KCl) and the substrate was first subjected to anodic potential cycling for electrochemical pretreatment. After rinsing, 0.1 cm^3 of a 10 mM ethanolic solution of AUDT was placed in the cell and the SAM was developed for 24 h at room temperature. 1,1'-Ferrocenedicarboxylic acid was dissolved in 1,4-dioxane to give a 0.1 M solution and was reacted with a slightly excess amount of CMCpTs for 2 h at RT. A 0.1 cm^3 portion of the solution was placed in the cell and was allowed to react with the AUDT SAM for 24 h at 25 °C. After rinsing, a 0.1 cm^3 portion of an aqueous 0.1 M CMCpTs solution was added to the cell and the surface-remaining carboxy groups were activated at 5 °C for 30 min. Finally, a 0.1 cm^3 portion of aqueous 20 μM 5'-amino ODN solution was added to the cell and a modified electrode with a DNA/ferrocene bilayer structure was obtained. *In-situ* hybridization was made by treating the electrode with an aqueous solution containing the complementary strand (20 μM) at 5 °C for 24 h.

CV measurements were made using an EG&G Co. Model 263A potentiostat. A Pt wire (ϕ 0.8 × 20 mm) and standard Ag/AgCl (sat. KCl) were used for counter and reference electrodes, respectively. Unless otherwise indicated, measurements were made in an aqueous 1.0 M KCl solution at 50 mV s^{-1} and room temperature, *ca.* 22 °C.

Results and Discussion

Immunosensor Application of DNA-Modified Electrode with DNA/Au Structure

Detailed studies of the electrode reactions of nucleobases and nucleotides have clarified that the redox reactions irreversibly occur under a high overpotential (*8*). Thus, the DNAs spread on the electrode surface significantly block the redox reactions of solution species. However, the reaction still occurs to a considerable extent on the DNA-modified electrode prepared by the procedure described above. The sterically bulky DNA double-strand with a stiff, rod-like structure is easy to form uncovered, bare sites upon adlayer formation and the faradaic reaction is expected to occur on the bare Au sites remaining in the DNA layer. This enables the biosensor applications of the DNA-modified electrode by taking advantage of the "ion channel concept"; the electrode adlayer functions as an "ion channel" which can be specifically "opened" by the analyte ion, leading to a permeability change in the electrochemically-detectable indicator ions (3). The development of work from our laboratory can be summarized as follows.

Cyclic voltammetric (CV) measurements for the $[Fe(CN)_6]^{4-/3-}$ redox couple using the DNA-modified electrode showed that the magnitude of the peak currents were markedly small compared to that measured on a bare Au electrode. However, the CV profile of the solution species on the DNA-modified electrode significantly changes upon adding DNA-binding substrates. For the case of quinacrine, which is a well-known antimalarial drug, the peak currents increased with the increasing concentration of quinacrine, showing almost a linear relationship with the concentration of the drug in the range of $10^{-7} — 5 \times 10^{-7}$ M which then became saturated over the concentration of 8×10^{-7} M (4). When the electrode responses were compared for a series of analyte molecules with different DNA-binding affinities, the selectivity order using the present sensor system seemed consistent with the binding affinity of the substrates with double-stranded DNA (6). Finally, the difference in selectivity diminished almost completely when the DNA-modified electrode was treated by a hot (70 °C) aqueous 6 M urea solution that produced the denatured form (single-stranded DNA) (6).

Figure 1 shows CVs of the $[Fe(CN)_6]^{4-/3-}$ redox couple with the bare and the DNA-modified Au electrode. The peak currents due to the reversible electrode reaction of $[Fe(CN)_6]^{4-/3-}$ on the bare electrode were suppressed to some extent by treatment with the thiol-modified DNA. Suppression of the electrode reaction was also associated with the subsequent treatment with ME (*vide infra*), but only slightly. When a serum protein, mouse IgM was added to the measurement solution (2×10^7 M), only a negligibly small electrode response was observed. In contrast, the electrode dramatically responded to the anti-DNA IgM which recognizes and binds to DNA. The CV response almost recovers if the electrode was treated with a 1 M KCl aqueous solution after the measurements. These results suggest that the DNA double-strands anchored on the electrode surface functions as the ligand molecule for the binding reaction of the antibody. In addition to the decrease in the CV peak current, the expected immunoreaction resulted in an increase in the peak separation. The former parameter is proportional to the effective area of the electrode while the latter is related to the heterogeneous electron-transfer rate constant. The expected immunoreaction affects both of the potentially distinct electrochemical parameters; binding of anti-DNA IgM (Mw 9×10^5) leads to steric hindrance

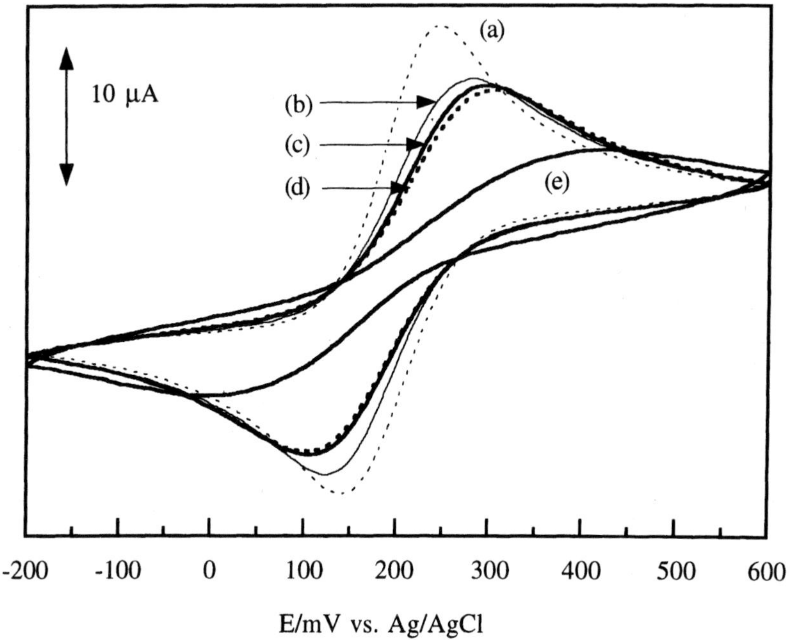

E/mV vs. Ag/AgCl

Figure 1. Cyclic voltammetric changes on the course of the electrode preparations and antibody responses: (a) bare Au electrode; (b) DNA-modified electrode, (c) after ME treatment, (d) in the presence of of 200 nM mouse IgM, (e) in the presence of 200 nM anti-DNA IgM.

which reduces the effective area of the electrode, and at the same time, limits the close contact of the redox indicator with the underlying electrode.

Treatment of the DNA-modified electrode with aqueous ME solution was essential to maintain the primary selectivity of the present immunosensor. If the DNA-modified electrode was used as prepared, the immunosensor also responded to other serum proteins, *e.g.*, mouse IgM, IgG, and bovine serum albumin. In most cases, this is due to the nonspecific adsorption of proteins onto the uncovered, bare Au site in the DNA adlayer as reported in the literature. We made an exploratory study using a variety of alkanethiols with the ω-functionality. As commonly reported, the hydroxyl terminus gave the most preferable results for the masking of the bare Au site. The blocking effect of the masking agent itself (Figure 1) should be considered since it makes the immunosensor application of the DNA-modified electrode difficult. We found that ME was best for such an application and was used for the electrode preparations throughout the present study.

CV responses to the anti-DNA IgM are shown in Figure 2. The anodic peak current decreased with increasing concentration of the antibody in the range of 10^{-9} — 5×10^{-8} M which then became saturated over the concentration of 7×10^{-8} M. The detection limit was around 5×10^{-10} M of antibody. The anti-DNA antibody has been used as a marker molecule of Systemic Lupus Erythematosus (SLE) which is a severe autoimmune disease (*9*). For the electrochemical detection of the anti-DNA antibody, a method combined with enzyme immunoassay (EIA) has already been reported (*10*). However, EIA requires an additional enzyme and seems to be somewhat complicated and time consuming. It can be concluded that the immunosensor described here has advantages in the much simpler and rapid operations. In addition, the present biosensor system should be potentially applicable for the detection of various DNA-binding proteins and the analyses of DNA-protein interactions.

Immunosensor Application of DNA-Modified Electrode with DNA/Fc/Au Structure

The creation of organized molecular membranes has been an increasingly important area in analytical chemistry. The methodologies of Langmuir-Blodgett and self-assembled monolayer (SAM) membranes are the most commonly used routes for organized film creations (11,12). We took advantage of the SAM formation from an ω-aminoalkanethiol to build a unique interfacial structure by subsequently derivatizing the amino terminus in the resulting monolayer. Derivatizations have also taken advantage of a coupling agent, CMC*p*Ts, to form more complex architectures through amide linkages.

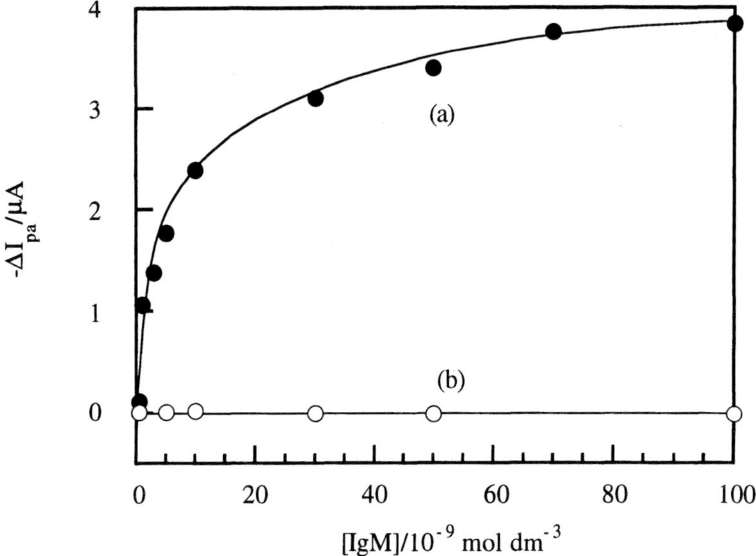

Figure 2. Electrode responses to anti-DNA IgM (a) and mouse IgM (b).

Scheme 2 shows the pathway for the preparation of the DNA-modified electrode with the DNA/ferrocene/Au structure. SAMs from AUDT was first developed on a Au electrode surface, then Fc was allowed to react with the amino terminus in the SAM to form an electroactive adlayer. The resulting Au electrode showed redox activity, typically 1.0×10^{-10} mol cm^{-2} of Fc as determined by the CV measurements. The Au electrode was subsequently treated with CMCpTs to activate the carboxy acid moieties remaining in the monolayer and a coupling reaction with an ODN having 5'-amino function (dT12) gave an adlayer with a DNA/Fc bilayer format. Finally, *in-situ* hybridization with the complementary ODN (5'-dA$_S$5-dA12-3') was made and a DNA-modified electrode with a DNA/Fc/Au interfacial structure was obtained. Examination of the substrate by FTIR reflection-absorption measurements confirmed that the surface phase formed reflected the intended adsorbate.

Upon DNA-immobilization, *ca.* an 80 % decrease in the electrochemical activity of the Fc-electrode was observed. As noted in the following section, this deactivation is not due to the disattachment of Fc from the surface, but is probably due to the effect of charge-compensation; the positive charges associated with the Fc oxidation require a flux of anionic charges (diffusion of counter ions) to maintain electroneutrality in the vicinity of the electrode surface, however, the polyanionic DNA located atop the Fc layer is expected to prevent flux from the bulk.

Figure 3 shows the CV responses to anti-DNA IgM. Interestingly, the CV peak currents increased with increasing concentration of the antibody. At present, we do not understand where the antibody response comes from. However, we speculate that it is related to the conformational changes in the DNA that are associated with protein binding in most cases. Structural changes that occur in the outer electrode adlayer would lessen the tight, phase-to-phase interaction with the underlying Fc layer. This leads to enhancement in the charge flux from the bulk making the electrode reaction feasible. Therefore we think that the present sensory system can operate without any electrochemically active indicator molecules.

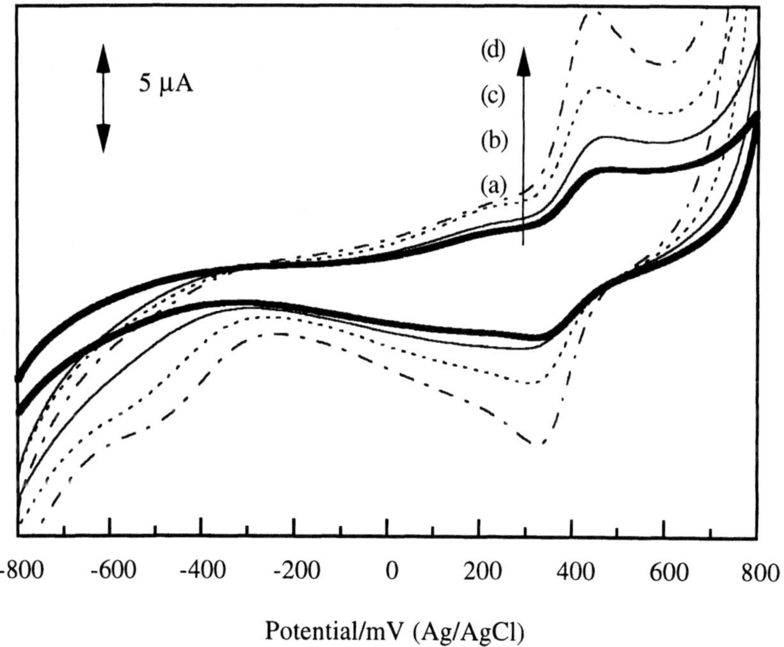

Figure 3. Cyclic voltammetric responses to (a) 0, (b) 1, (c) 5, (d) 10 nM of anti-DNA antibody on the DNA/Fc/Au electrode.

Plots of the anodic peak current vs. IgM concentration are shown in Figure 4. As seen in this figure, the present immunosensor should be applicable for the determination of the antibody in the range of 10^{-9} — 10^{-8} M. The sensor response was reversible; the CV peak current was recovered if the electrode was treated with 1 M KCl aqueous solution without the antibody. Repeated use (20 times) within 24 h gave reproducible responses toward the antibody. However, the modified electrode did not appear to be very stable since the response was lost a few days later. A detailed characterization study of the immobilization chemistry, understanding of the electrode response and influences of possible coexisting molecules are now on going in our laboratory and will be reported elsewhere.

Acknowledgments

This work was partially supported by a Grant-in-Aid for Scientific Research (No. 11650831) from the Ministry of Education, Science, Sports, and Culture of Japan. Financial support from The Ogasawara Foundation for the Promotion of Science and Engineering (K. N.) is gratefully acknowledged.

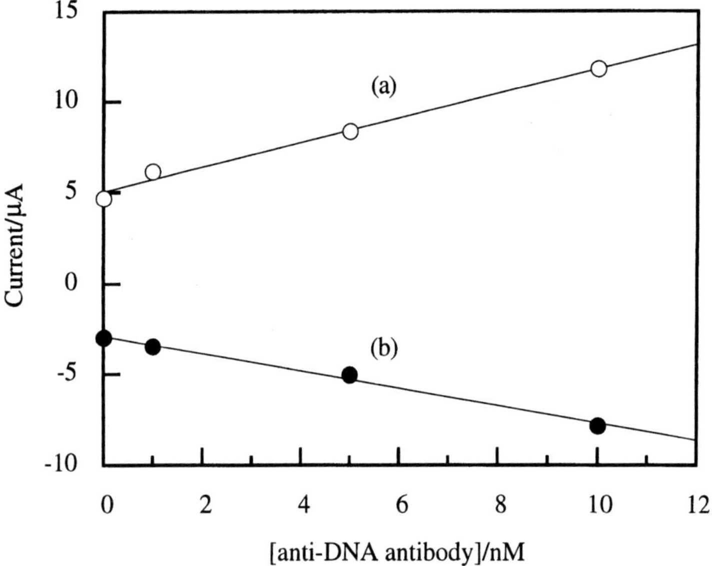

Figure 4. Electrode response in CV anodic (a) and cathodic (b) peak currents to anti-DNA antibody.

References

1. *Genomes*; Brown, T. A. (Japanese Edition by Muramatsu, M.); Medical Sciences International, Ltd. : Tokyo, 2000.
2. *Bioorganic Chemistry: Nucleic Acids*; Hecht, S. M. Ed.; Oxford Press: New York, 1996; p 324.
3. Sugawara, M.; Kojima, K.; Sazawa, H.; Umezawa, Y. *Anal. Chem.* **1987**, *59*, 2842.
4. Maeda, M.; Mitsuhashi, Y.; Nakano, K.; Takagi, M. *Anal. Sci.* **1992**, *8*, 83.
5. *Redox Mechanisms and Interfacial Properies of Molecules of Biological Importance*; Schultz, F. A.; Taniguchi, I. Eds.; The Elctrochemical Society, Pennington, 1993; p 423.
6. *ACS Symposium Series 690 Polymers in Sensors. Theory and Practice*; Akmal, N.; Usmani, A. M., Eds; American Chemical Society: Washington, DC, 1998; p 34.
7. Tazaki, M.; Yakata, K.; Uemura, T.; Nakano, K.; Sakai, M.; Yonemitsu, T.; Nagahama, S. *Sulfur Lett.* **2000**, *23*, 259.
8. *Electrochemistry of Biological Molecules*; Dryhurst, G.; Academic Press: New York, 1977; p 269.
9. Hashimoto, H.; Tsuda, H.; Takasaki, Y.; Fujimaki, N.; Suzuki, M.; Shimokawa, Y. *Scand. J. Rheum.*, **1983**, *10*, 727.
10. Babkina, S. S.; Medyantseva, E. P.; Budnikov, H. C.; Tyshlek, M. P. *Anal. Chem.*, **1996**, *68*, 3827.
11. Porter, M. D.; Bright, T. B.; Allara, D. L.; Chidsey, C. E. D. *J. Am. Chem. Soc.* **1987**, *109*, 3559.
12. Ulman, A. *Chem. Rev.* **1996**, *96*, 1533.

Optical Sensing and Processing

Chapter 6

Digital Optical Chemistry: A Novel System for the Rapid Fabrication of Custom Oligonucleotide Arrays

Kevin J. Luebke[1], Robert P. Balog[1,2], David Mittelman[1,2], and Harold R. Garner[1,2]

[1]Center for Biomedical Inventions, University of Texas Southwestern Medical Center, Dallas, TX 75390–8573
[2]McDermott Center for Human Growth and Development, University of Texas Southwestern Medical Center, Dallas, TX 75390–8591

We describe the design and initial characterization of an automated DNA array synthesizer that uses a Digital Micromirror Device to control light-directed chemistry. Ultraviolet light reflected from the micromirrors is projected to the reactive surface of a glass substrate to effect the photodeprotection step of oligonucleotide synthesis, and reagents for the subsequent coupling are delivered to the surface. Repetitive cycles of illumination and coupling build an array of probe oligonucleotides on the surface. The probe oligonucleotides are synthesized with stepwise yield greater than 95% and can discriminate against single-nucleotide mismatches in binding to complementary DNA. Resolution of the projected light is high enough to synthesize more than 49,000 discrete probe features in parallel. This method of array fabrication offers benefits in flexibility and economy over other common methods of array fabrication.

New methods for the fabrication of nucleic acid arrays are needed to realize the potential of these powerful tools of biological inquiry. Sometimes called DNA chips, arrays of DNA probes immobilized in discrete regions on a surface allow the parallel analysis of nucleic acid sequences in complex mixtures of DNA and RNA. Each area, or feature, contains a different probe sequence, and each is the site of an independent binding assay. The remarkable specificity and predictability of molecular recognition between a nucleic acid polymer and its complementary sequence allows automatic selection of probes and interpretation of the binding at each feature (*1*). Arrays with hundreds of thousands of microscopic features can be created and analyzed.

Nucleic acid arrays are powerful, because they provide rapid access to large amounts of genotypic and phenotypic information. They are usually used in two general types of assays. In one of these assays, sequence information is sought (*2-5*). This method works best to determine variations from a known reference sequence, so it is often termed re-sequencing. It distinguishes between different sequences by taking advantage of the sensitivity of binding to single-nucleotide mismatches. The other type of assay uses binding to probes to quantitate DNA or RNA of particular sequences. For example, the relative amounts of thousands of messenger RNA species in a cell or tissue can be measured simultaneously, providing an instantaneous profile of its metabolic state (*6-8*).

A variety of methods have been developed for the fabrication of nucleic acid arrays. In some, each DNA probe is prepared separately and subsequently arrayed on the surface by automated deposition (*8, 9*). This method allows facile characterization of the probe molecules to be deposited, but the costs in labor, materials and time of synthesizing and manipulating thousands of separate probe molecules are high. Also, the density of features per unit area is limited by the resolution with which probe solutions can be deposited.

In other methods, the steps of DNA synthesis and arraying are performed simultaneously, each DNA probe being synthesized from monomers *in situ* at the correct position on the array (*10*). The general sequence of chemical reactions is outlined in Figure 1. A hydroxyl group on the surface or at the 5' terminus of a growing oligonucleotide reacts with the next monomer, which is introduced as a phosphoramidite. The 5' hydroxyl of this phosphoramidite is chemically blocked with a protective group to ensure that only one monomer unit is added to any nascent oligomer. The new linkage is oxidized to the stable phosphate triester before the 5' protective group is removed and another round of monomer addition is initiated. When all of the monomers have been added, the protective groups on the phosphates and the heterocyclic bases are removed with mild base.

Fodor and co-workers realized that a *photolabile* protective group can be used to block the 5' hydroxyl groups of the monomers, allowing the synthesis of different sequences on a surface with the spatial resolution of illumination (*5, 11*). The method is illustrated in Figure 2. The surface of the array is covered with photolabile protected hydroxyl groups. Portions of the surface to which the next monomer are to be coupled are illuminated with light in the near ultra-violet region of the spectrum, which removes the photolabile moieties, revealing free hydroxyls. The entire surface is flooded with the desired phosphoramidite, but only the patches that have been deprotected with light react. Thus, by repetitive cycles of this process, a different, known sequence can be built in any

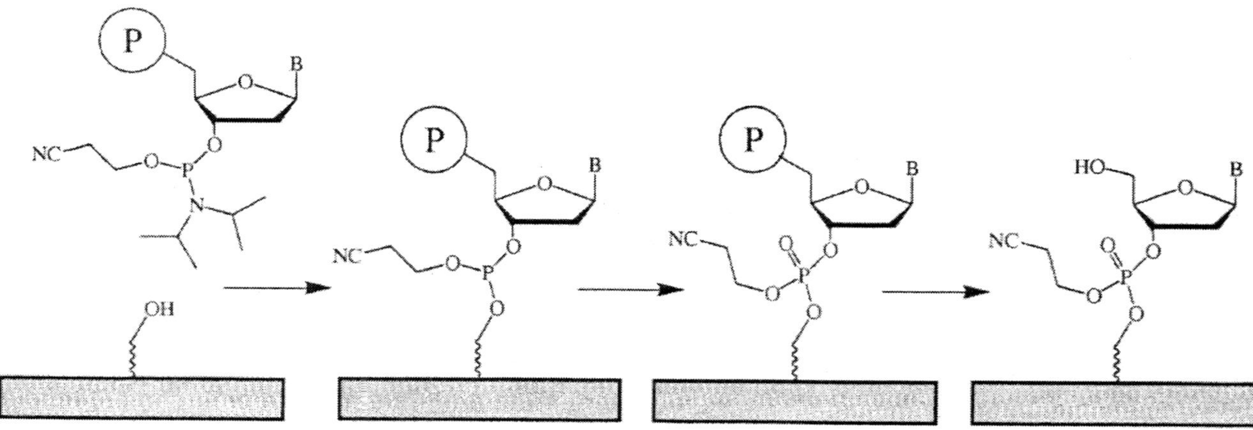

Figure 1. The sequence of chemical reactions for the synthesis of oligonucleotide probes on hydroxylated glass substrates. The heterocyclic bases, B, are suitably protected. The 5' hydroxyl protective group, P, is removed prior to each cycle of monomer addition.

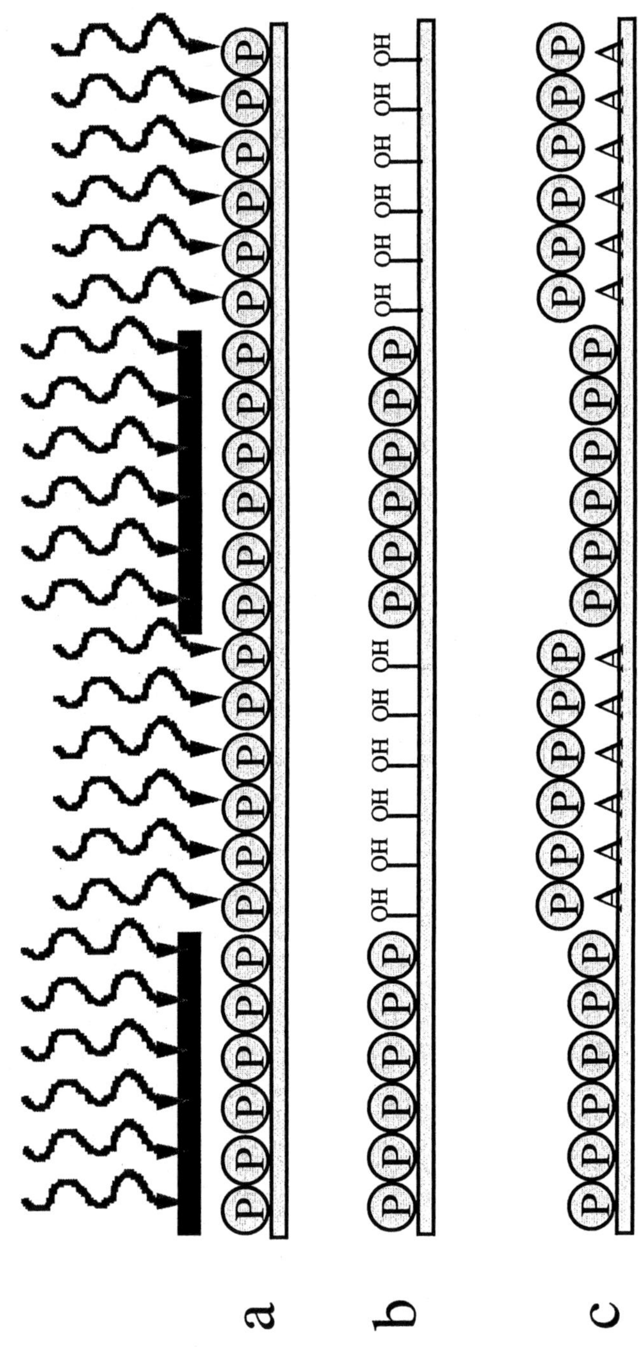

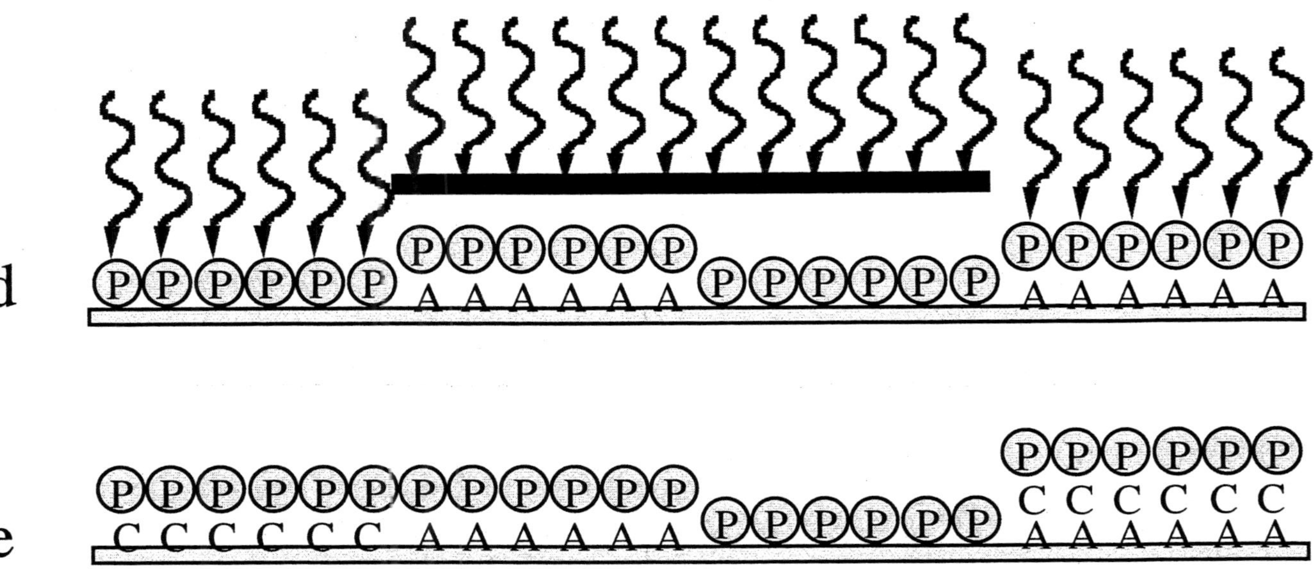

Figure 2. Procedure for light-directed synthesis of oligonucleotide arrays. a) Selective irradiation of regions to which the next base will be coupled. b) Surface after irradiation with selected hydroxyls deblocked. c) Coupling of the first base, A, followed by d) deprotection of the regions to be coupled next, and e) another coupling step, this time to C.

region of the surface that can be discretely illuminated. Affymetrix has developed this method into a commercial process in which they use physical photolithographic masks as barriers to light in the regions that are to remain protected in each synthetic cycle. This photolithographic process allows mass production of oligonucleotide arrays comprising hundreds of thousands up to a million features within a square centimeter.

The use of physical masks for the production of arrays has the drawback, however, that it limits the experimental flexibility of the individual researcher. Because of the time required to fabricate the masks, the materials costs, and the infrastructure requirements of the mask-based photolithographic process, a new array with a set of oligonucleotide probes that are custom-designed for a given project requires months to create and costs thousands of dollars. A rapid, economical route to high feature-density arrays would benefit the project-oriented researcher who needs a small number of arrays of a novel design to answer a particular question.

The need for physical masks is obviated by an alternative method of projecting patterned ultraviolet light with high resolution and contrast. A projection technology developed by Texas Instruments embodies this alternative (*12-14*). The technology uses a Digital Micromirror Device (DMD), a 1024 x 768 array of microscopic aluminum mirrors, each 16 µm on a side, machined onto the surface of a chip. Each mirror pivots on a central post and can be individually controlled to tilt +10° or −10°. Light reflected from selected mirrors can thus be projected to create an image in which each mirror represents a single pixel. The configuration of the mirrors and thus the image can be altered under computer control on the time-scale of microseconds.

Our application of this projection technology to the fabrication of nucleic acid arrays is shown schematically in Figure 3 (*13*). The DMD is uniformly illuminated with ultraviolet light and configured to project an image that illuminates the desired regions of a surface. The surface in this case is a glass microscope slide that has been chemically coated with protected hydroxyl groups. It is held in a closed chamber to which reagents for DNA synthesis are delivered. Delivery of reagents and display of each image are automatically coordinated by computer. We call this system for array synthesis Digital Optical Chemistry (DOC).

Design of the System

Optical System

We have developed an optical system with the functions of illuminating a DMD and projecting the reflected image onto the reactive glass surface. Illumination of the DMD must be accomplished with sufficient light intensity at

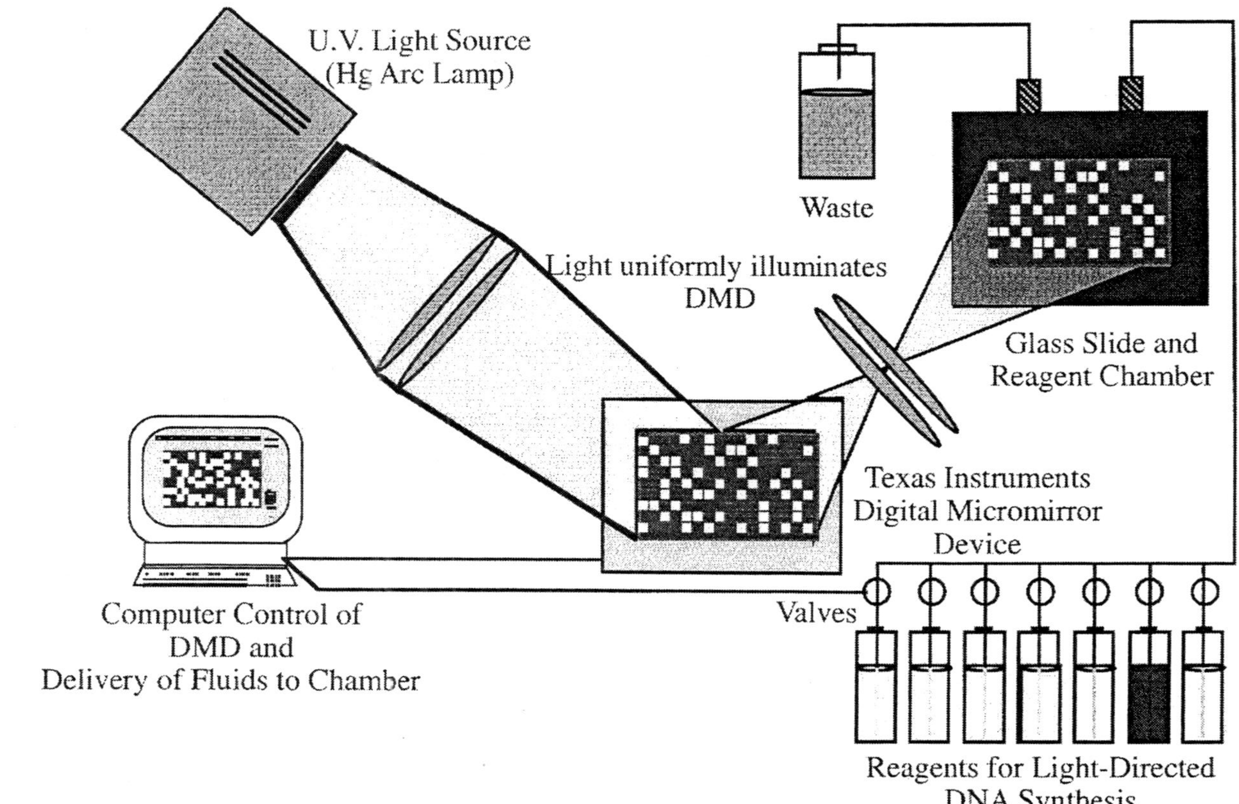

Figure 3. Schematic diagram of the Digital Optical Chemistry system.

appropriate wavelengths that the projected light provides efficient deprotection in a reasonable amount of time. The protective group we are using, ((α-methyl-2-nitropiperonyl)oxy)carbonyl (MeNPOC), is effectively cleaved with light at 365 nm (*15*). Thus, we are illuminating the DMD with a 1 kW Hg arc lamp, which has an intense emission band at 365 nm. Because only the light in this band is needed to effect the desired chemistry, other wavelengths are removed, and correction of achromatic aberrations is unnecessary. A cold mirror filters out wavelengths above 400 nm and a 365 nm interference filter, with a transmission peak width at half maximum of 2 nm, selects for the desired wavelength. Removal of wavelengths below 340 nm is necessary to avoid light-induced damage to the DNA being synthesized (*16*).

Uniformity of illumination of the DMD is also beneficial. Contrast between illuminated and dark regions of the image is imperfect, so there exists an optimum amount of energy to impart to the reactive surface to achieve maximum fidelity of probe synthesis (*vide infra*). Non-uniformity of illumination will result in the inconvenience of a different optimal deprotection time at different positions in the image. Light from the Hg lamp is collimated by a silica lens system, and the collimated light beam is reflected from a front surface mirror to a spherical mirror. The spherical mirror reflects the beam of light to the DMD at an angle of 20° off normal, resulting in a reflected beam that is normal to the plane of the projection lens.

The image reflected by the DMD must be projected onto the reactive surface with high resolution. Ideally, a single mirror on the DMD will address a discrete feature on the array, affording 786,432 (1024 x 768) different oliognucleotide probes. The mirrors are 16 μm on each side with 1 μm spacing between mirrors, an appropriate size for an individual feature in a high-density oligonucleotide array. No reduction or enlargement of the image is necessary. Projection of a 1:1 image results in an array with dimensions of 17 mm x 13 mm and has the benefit that no correction of spherical aberration is required.

To faithfully synthesize the desired oligonucleotide sequence in each feature, the image reflected by the DMD must also be projected with high contrast. That is, the ratio of light intensity striking the reactive surface in illuminated areas to the light intensity striking the nominally dark areas must be high. The rate of protective group photocleavage is directly proportional to the light power delivered per unit area (*15*). Thus, light striking the reactive surface in features that are nominally dark will result in deprotection in those features and subsequent insertion of an inappropriate base. Reducing the deprotection time decreases the frequency of these insertion errors at the cost of incomplete deprotection in the features where coupling of the next base is desired. The sum of the resulting deletion and insertion errors is minimized at an optimal deprotection time that depends upon the contrast ratio. The primary effect of these synthesis errors is to raise the lower limit of detection for an analyte by decreasing the surface density of functional, correct-sequence probes in any given feature.

The contrast of the image is limited partly by the DMD itself. For example, light scattered from the edges of the mirrors and the material behind the mirrors limits the contrast ratio. Care must be taken, though, to avoid degradation of the contrast when relaying the image to the reactive surface. To project a high-contrast, high-resolution image of the DMD, we have used a custom six-lens element with anti-reflective coating tuned to 365 nm. A light dump consisting of optical felt absorbs the rejected light.

Digital Image Generation

Images for photodeprotection can be generated in sets of four, one image for each of the four bases, one set for each base added to an oligonucleotide on the array. Thus, an array of probes each with 25 bases would require 100 masks. Nevertheless, fewer deprotection and coupling steps can be used by deprotection according to the next base needed in a feature, irrespective of the position of the base from the end of the growing chain. The four bases are repetitively introduced in the same order. For example, if phosphoramidites are introduced in the order ATCG, and a probe's first two bases are 3'-AC-5', both the A and C can be added within three coupling steps. This method differs from the position-dependent method, because seven coupling steps would have been required to add the first two bases, cycling through addition of all four phosphoramidites before the second base could be added. This method reduces the number of images necessary to make an array by >20%. To determine the optimum coupling sequence of the four bases, synthesis of each array design is simulated using all 24 possible sequence orders. Decreasing the number of coupling steps decreases the amount of time and the reagents used for an array, and it also increases the fidelity of probe synthesis. Insertion and deletion errors made in each deprotection step are minimized by minimizing the number of steps.

Instrument Control Software

To maximize the flexibility of the DOC system, a simple interpreted scripting language was developed. The script language allows users to customize the entire chip synthesis process. Natural language syntax was chosen to make the script easy to learn and to help retain its usefulness to both the user and the computer. Because the script language is flexible, it leaves room for system improvements. For example, the software does not assume a fixed number of valves or even that the system will be used to build oligonucleotides.

The same software could be used in conjunction with different chemicals, protocols, and hardware to build other molecular library arrays.

Fluidic System

Delivery of reagents to the substrate, as in conventional DNA synthesizers, occurs in a closed system. The bottles with individual reagents are pressurized with dry Argon, and the reagents are directed to the substrate via a network of Teflon® valves, connectors, and lines. Mixing of different reagents, when necessary, is accomplished in a common line by pulsing the valves between the open and closed state on a time scale of 20 milliseconds. Small alternating aliquots of the reagents are created and mix at their interfaces. To prevent unwanted cross contamination between reagents in the common line, each reagent line is attached to a dedicated acetonitrile line. This feature allows routine washing of each reagent line and the common line between each DNA synthesis step.

Characterization of the System

Time Course of Photodeprotection

To find the optimal deprotection time, which minimizes insertion and deletion errors for a given contrast ratio, the rate of photodeprotection for the system must be determined. We have directly measured the rate of reaction in our system using modifications of methods that have been described previously (Figure 4) (*15*). A MeNPOC-protected nucleoside phosphoramidite was coupled to a hydroxy-alkylated glass substrate mounted in the instrument reaction chamber. Eight rectangular regions were illuminated with the DMD, each for a different, known amount of time. Finally, the relative density of hydroxyls in each region from which MeNPOC had been photolyzed was probed by coupling with a phosphoramidite derivative of the fluorescent dye Cy3. To ensure that the fluorescence intensity was proportional to the density of free hydroxyls and not subject to quenching due to interaction between proximal fluorophores on the surface, the Cy3 phosphoramidite was diluted to a mole fraction of 0.01 with an equally reactive, but non-fluorescent, phosphoramidite, 5'-dimethoxytrityl-thymidine-3'-*O*-(β-cyanoethyl)-*N*,*N*-diisopropylphosphoramidite.

The fluorescence image of the surface (Figure 4a) was acquired with a confocal laser scanner, and the fluorescence intensity for each illuminated region was plotted as a function of illumination time (Figure 4b). The fluorescence intensity increased as a function of time and the data were fit to a first-order exponential function from which the rate constant was determined. The measured rate constant was 1.73 min.$^{-1}$, corresponding to a photolysis half-life of 24 seconds. From data reported by McGall and co-workers (*15*) relating MeNPOC photolysis rates to light power per unit area, we can calculate that the power delivered by our system to the reactive surface is 16 mW/cm^2.

The total synthesis time for an array is determined by the time required for photolysis prior to each coupling step and by the time required for each coupling cycle subsequent to deprotection. Deprotection for 8 half-lives contributes 3 minutes to the total time for each cycle and subsequent steps require 14 minutes, resulting in a time of 17 minutes for each coupling cycle. Synthesis of an array of 20-nucleotide probes typically requires 48 deprotection and coupling cycles. Thus the total time for synthesis of an array on our system is approximately 14 hours.

Stepwise synthesis yield

A high surface density of probe molecules in each feature is desirable. Below densities at which probe oligonucleotides interact with each other or sterically restrict binding of analyte (*1*), the amount of analyte that hybridizes (and hence the signal) at a given concentration increases with increasing density of probes. The density of full-length probes is the product of the initial density of MeNPOC-protected hydroxyl groups and the fractional yield of the synthesis. The overall synthetic yield equals the synthetic yield for each step raised to the power of the number of steps in the synthesis. For example the full-length yield of a 25 mer with an average stepwise yield of 98% is $0.98^{25}=0.60$. Small differences in stepwise yield result in large differences in the full-length yield of probe. Two aspects of the performance of the system determine the stepwise yield: the coupling yield, which depends upon the appropriate delivery of reagents to the reaction chamber by the fluidic system, and the deprotection yield, which depends upon efficient photolysis of the protective group. Unwanted side-reactions are known to accompany photolytic deprotection of MeNPOC-protected hydroxyls (*15*). The products of these side-reactions are termini that can neither couple to a phosphoramidite nor be correctly deprotected.

The stepwise synthesis yield in our system was measured by parallel synthesis of oligomers of different lengths. The experimental method was similar to that used to determine the photolysis rate. A monomer was coupled to a substrate, and eight separate regions were illuminated to effect deprotection. A long illumination time of 12 half-lives was used to ensure maximum efficiency of deprotection. A second base was coupled to each deprotected region and remaining unreacted hydroxyls were capped by acetylation. Next, seven of the

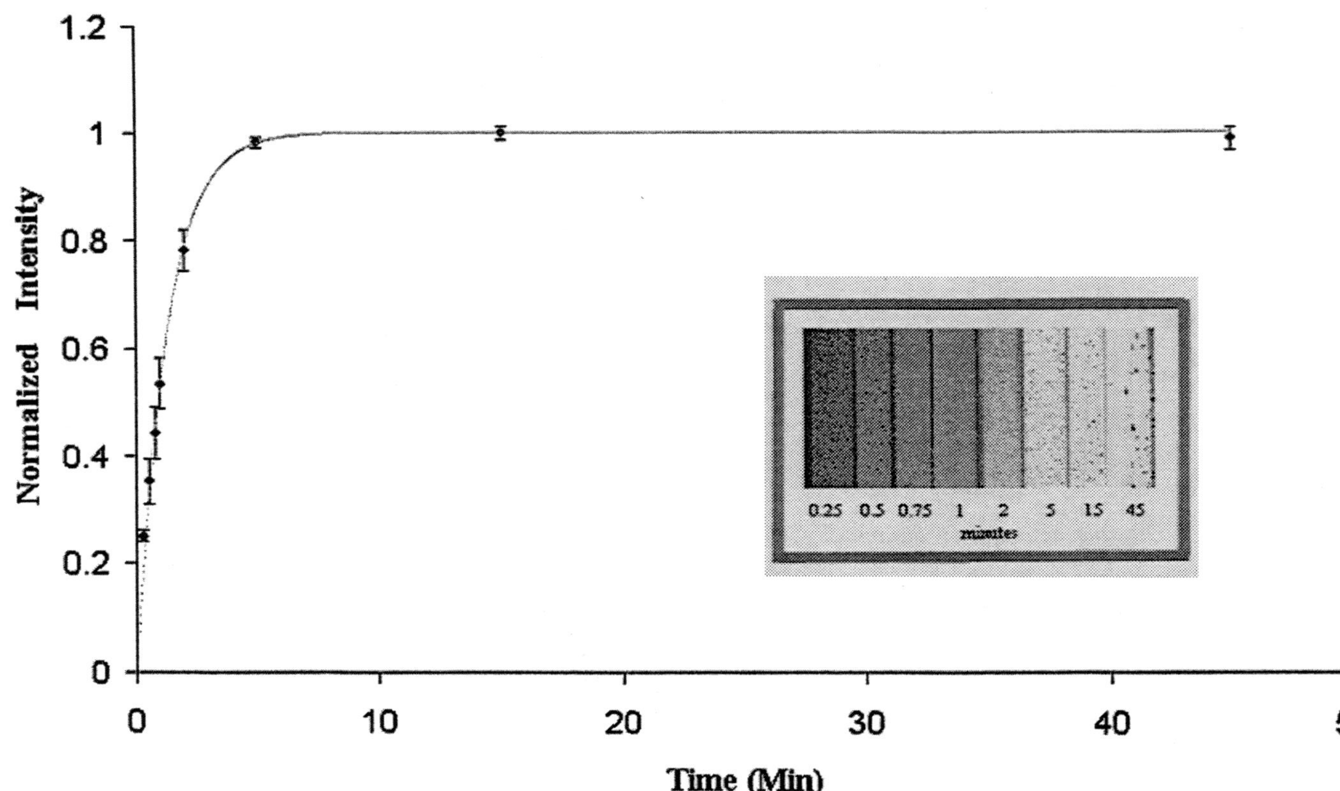

Time Course
Normalized Intensity
Time (Min)
0.25 0.5 0.75 1 2 5 15 45
minutes

Figure 4. Time course of photodeprotection. The glass substrate was coated with bis(hydroxyethyl)aminopropylsilane and uniformly coupled with MeNPOC-protected thymidine phosphoramidite. The inset shows the fluorescence image resulting from illumination for the times indicated, followed by probing for deblocked hydroxyls with Cy3 phosphoramidite, and scanning the fluorescence intensity. Fluorescence intensity versus illumination time is plotted. The curve is the best fit to the data for a first-order exponential function.

regions were photodeprotected, coupled, and capped, leaving only two bases per oligomer in one region, and three in the others. In each subsequent cycle, one fewer region was deprotected, coupled and capped, resulting in oligomers of eight different lengths, two to nine. Finally, oligomers in all eight regions were deprotected by illumination.

The relative density of free hydroxyls in each region after deprotection indicates the relative synthetic yield of full-length oligomer in that region. Truncations of the synthesis prior to the final deprotection result in strand termini that cannot be photolyzed to yield hydroxyls. Surface hydroxyls can again be probed by coupling with a diluted solution of Cy3 phosphoramidite and quantitating fluorescence. The ratio of fluorescence between regions of consecutive oligomer length is the synthetic yield for that coupling step. The mean stepwise yield for a synthesis of oligomers of dA was measured to be 95.9%. A similar stepwise yield of 96.3% was measured for the synthesis of a 25mer with sequence 5'-dTd(GCAT)$_6$-3'. This yield is comparable to that reported by others using the same chemistry (*14,15*). More than 98% of surface hydroxyls are blocked from reaction with Cy3 phosphoramidite by automated coupling in our system, indicating that some decrease in yield is sustained due to side reactions in the photolytic deprotection.

Image Quality

We have quantitated the resolution and contrast of the optical system by photochemically patterning the reactivity of a substrate to fluorescent phosphoramidite. MeNPOC-protected thymidine phosphoramidite was coupled uniformly to a hydroxylated glass substrate. The substrate was next illuminated with an image that consisted of a series of checkerboard patterns. The squares in each checkerboard were generated by a different number of mirrors, 625, 100, 16, and 4 total mirrors per square (25, 10, 4, and 2 mirrors on a side, respectively). Following illumination, the pattern of deprotection was visualized by treating the substrate with dilute Cy3 phosphoramidite and fluorescence scanning. The resulting image is shown in Figure 5. Individual squares generated by 16 (4 x 4) mirrors were clearly resolved. Visualization of squares generated by 4 (2 x 2) mirrors is limited by the 10 μm resolution of the scanner used, the features being 34 μm x 34 μm. Nevertheless, a checkerboard pattern can be discerned. Thus, a requirement of 16 mirrors per feature is an upper limit, and the lower limit of the possible number of features per array is 49,000. Determination of the upper limit of feature number relies upon a higher resolution tool for visualization.

To determine the contrast, the fluorescence signal was quantitated in both light and dark areas. The optical contrast ratio was calculated from the ratio of the background-corrected signals in light and dark areas, the known rate

Figure 5. Fluorescence image from an experiment to test resolution and contrast. The squares in each checkerboard were generated by a different number of mirrors, 625, 100, 16, and 4 total mirrors per square (25, 10, 4, and 2 mirrors on a side, respectively). See text for experimental details.

constant for the photolysis reaction in the illuminated regions, and the assumption that the reaction follows first-order kinetics with a rate constant that is directly proportional to the light power per unit area (*15*). The contrast ratio between the largest light and dark features was 596:1. With this contrast ratio, the minimum number of errors is created when the deprotection is for 7.7 half-lives, leaving 0.5% protected to become deletion errors in each coupling cycle and incorrectly deprotecting 0.9% to create insertion errors in each coupling cycle. These errors will be distributed randomly among the oligomers at each step such that in the synthesis of an array of 20-nucleotide probes, using 48 coupling cycles, 31% of all of the full-length probes will contain one or more insertions or deletions. These erroneous oligomers contribute little or nothing to the observed hybridization signal, so the primary effect of the synthesis errors is to decrease the maximum signal that can be generated.

Sequence Specific Hybridization to Arrays

As an initial test of the performance of arrays generated with DOC, we created an array for the sequence analysis of a short piece of DNA. The model analyte was a 25-nucleotide oligomer that was chemically synthesized with a fluorescent Cy3 label at its 5' end. The sequence of this oligomer was from the HIV genome. The array comprised 144 features, each containing a 20-nucleotide probe. In addition to features in which the probes were the perfect complement to 20 nucleotides of the analyte, features were included with probes that had the same sequence composition as the perfect match probes but a scrambled order. Probes that corresponded to each single-nucleotide mismatch within 20 nucleotides of the analyte were also included.

The model analyte was allowed to hybridize to the array, and after washing the array, binding was visualized by fluorescence scanning. A part of the fluorescence image is shown in Figure 6. The bottom checkerboard of 16 features contains probes that are the perfect complement to the analyte (bright features) and scrambled sequence probes (dark features). The difference in fluorescence signal between these two sets of features indicates sequence-specific binding of the analyte to the array. Sensitivity of binding to single-nucleotide mismatches is seen in the upper block of sixteen features. In each of the four columns of this block, a different base in the analyte is probed, bases at positions 9, 10, 11, and 12 (5'-GCAT-3') of the labeled oligomer from right to left. Each base in these probes is the complement to the analyte except at the position being queried, where the base is substituted with each of the four bases in the four rows of the block.

In each column, the brightest feature is that which has the perfect match probe, indicating single-nucleotide discrimination in binding of analyte to the array. The sequence of the four central bases can be read directly from the array.

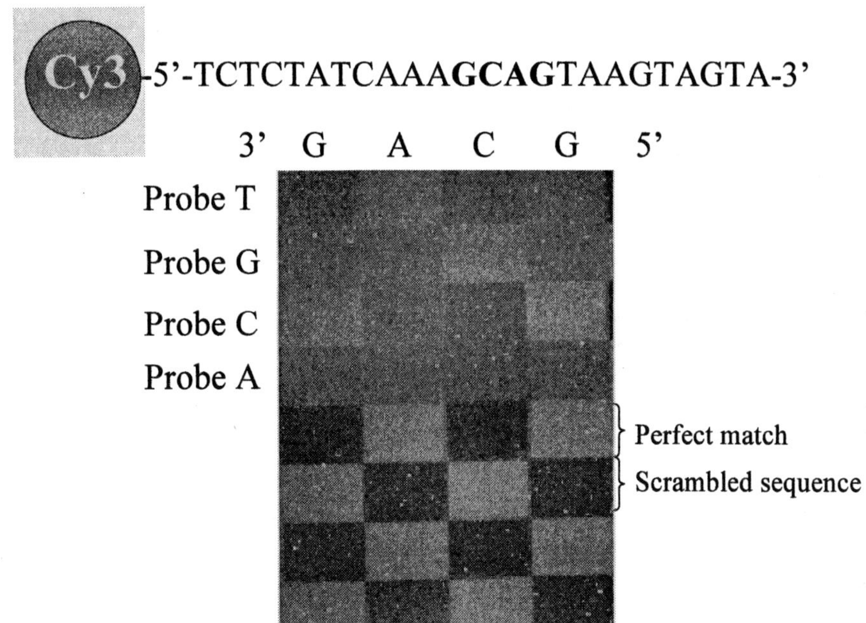

Perfect match probe: 5'ACTACTTACTGCTTTGATAT3'
Scrambled sequence probe: 5'TAATTAATGTACCCGTTACA

Figure 6. Sequence-specific binding to an array synthesized using DOC. The nucleotides listed in boldface type in the perfect match probe were varied in the upper block of sixteen features to probe the identities of the bases at complementary positions in the labeled analyte.

Taking the signal in the scrambled sequence features to be background, discrimination between correct sequence probe and single nucleotide mismatches at these positions is by a factor of 1.9 to 2.5. Others have reported that the highest sensitivity to mismatches is at centrally located positions within a probe oligonucleotide, and consistent with that observation, bases at positions outside the central four shown could not be unambiguously determined by this hybridization analysis. In a typical re-sequencing analysis with oligonucleotide arrays (*2-5*), each nucleotide is queried with a set of probes in which the complement to the queried base is at or near the center of the probes. Thus, the specificity of the array produced on the DOC system is sufficient for sequence analysis.

Summary and Future Prospects

Using a Texas Instruments Digital Micromirror Device to pattern light-directed synthesis of DNA, we have created an automated system that can synthesize a functional nucleic acid array in less than one day, starting from a novel design. The resolution of the projection system is sufficient to create at least 49,000 features in an area of two square centimeters. We have demonstrated DNA sequence analysis with an array made in this manner, confirming that standard re-sequencing analyses are enabled. Furthermore, we have found synthesis efficiency and fidelity to be comparable to those resulting from the mask-based photolithographic approach to array synthesis, suggesting that arrays made with this system will be useful for the entire repertoire of oligonucleotide array applications, including transcript profiling.

A principal benefit of the DOC system is the ability to proceed quickly from design of a new array of probes to a functioning array. Starting from a new list of 20-nucleotide probes to be synthesized, an array can be produced, ready to use, in less than 24 hours. It does not require the time or the manufacturing infrastructure needed to fabricate a set of physical masks as in the traditional photolithographic approach. Unlike the approach in which probes are synthesized separately and deposited to the array, the time required does not depend upon the number of different probes to be included, because as many probes as allowed by the resolution of the light projection, at least 49,000, can be made in parallel. This degree of parallelism is orders of magnitude higher than afforded by the highest throughput systems for oligonucleotide synthesis, template DNA amplification, or DNA purification currently available. Thus, with the DOC system, the number of probes (and hence the number of analytical targets) that can be investigated per unit time is high.

Another benefit of the DOC system is economy. Manufacture of physical masks is prohibitively expensive for an experimental array design of which only a few arrays might be used. Similarly, the conventional synthesis of a large number of oligonucleotides or PCR amplification of larger probes is a large expense that must be paid prior to the generation of a single array. These other methods benefit from an economy of scale that makes them best suited to fabrication of array designs for which there is certainty of a persistent need. Nevertheless, the high "setup" costs of these methods prohibit their application to the fabrication of a small number of custom arrays. Fabrication of arrays with DOC does not suffer from these initial costs. The cost of each array is approximately that of the reagents to make the array. The costs of the instrument and the space it occupies are relatively small, comparable to those for other machines that perform automated synthesis of biomolecules.

A number of enhancements to the current system are planned. Though DOC is not designed for mass production of arrays, increased throughput is nonetheless desirable, and several of the planned enhancements are aimed at that goal. For example, multiple arrays can be synthesized in parallel with a single DMD by rotating them in and out of the light path with each coupling cycle. A larger number of features per array is also desirable, and Digital Micromirror Devices are now available from Texas Instruments with 1.3 million mirrors (1280 x 1024). Installation of such a device will increase the number of features possible by a factor of 1.7.

Fabrication of DNA arrays with DOC allows greater flexibility than is available with generic arrays that are mass-produced by the traditional photolithographic approach. In addition, it affords a higher density of features than other custom fabrication approaches like deposition of nucleic acids and delivery of reagents for *in situ* synthesis with ink jets (*17*). The speed and economy with which new arrays of high feature density can be generated with DOC will be valuable to the researcher who wants to converge quickly on arrays of probes that best address particular questions.

Acknowledgments

We thank Dr. Glenn McGall of Affymetrix for helpful advice and generous donations of phosphoramidites. We also thank Dr. Eric Mayrand of PEBiosystems for helpful conversations. This work was supported by the P. O. B. Montgomery Chair Fund and the National Institutes of Health (CA81656-01). A McDermott Center Human Genomics training grant supported RPB.

References

1.   Southern, E.; Mir, K.; Shchepinov, M. *Nature Genetics Supplement* **1999**, *21*, 5-9.
2.   Wang, D. G.; et al. *Science* **1998**, *280*, 1077-1082.

3. Chee, M.; Yang, R.; Hubbell, E.; Berno, A.; Huang, X. C.; Stern, D.; Winkler, J.; Lockhart, D. J.;
Morris, M. S.; Fodor, S. P. A. *Science* **1996**, *274*, 610-614.
4. Hacia, J. G.; Brody, L. C.; Chee, M. S.; Fodor, S. P. A. *Nature Genetics* **1996**, *114*, 441-447.
5. Pease, A. C.; Solas, D.; Sullivan, E. J.; Cronin, M. T.; Holmes, C. P.; Fodor, S. P. A. *Proc. Natl. Acad. Sci. USA* **1994**, *91*, 5022-5026.
6. Rosania, G. R.; Chang, Y. T.; Perez, O.; Sutherlin, D.; Dong, H.; Lockhart, D. J.; Schultz, P. G. *Nature Biotechnology* **2000**, *18*, 304-308.
7. Wodicka, L.; Dong, H.; Mittman, M.; Ho, H.-H.; Lockhart, D. J. *Nature Biotechnology* **1997**, *15*, 1359-1367.
8. Duggan, D. J.; Bittner, M.; Chen, Y.; Meltzer, P, Trent, J. M. *Nature Genetics Supplement* **1999**, *21*, 10-14.
9. Guo, Z.; Guilfoyle, R. A.; Thiel, A. J.; Wang, W.; Smith, L. *Nucleic Acids Res.* **1994**, *22*, 5456-5461.
10. Maskos, U.; Southern, E. M. *Nucleic Acids Res.* **1992**, *20*, 1679-1684.
11. Fodor, S. P. A.; Read, J. L.; Pirrung, M. C.; Stryer, L.; Lu, A. T.; Solas, D. *Science* **1991**, *251*, 767-773.
12. Sampsell, J. B. *J. Vac. Sci. Technol.* **1994**, *B12*, 3242-3246.
13. Jaklevic, J. M.; Garner, H. R.; Miller, G. A. *Annu. Rev. Biomed. Eng.* **1999**, *1*, 649-678.
14. Singh-Gasson, S.; Green, R. D.; Yue, Y.; Nelson, C.; Blattner, F.; Sussman, M. R.; Cerrina, F. *Nature Biotechnology* **1999**, *17*, 974-978.
15. McGall, G. H.; Barone, A. D.; Diggelman, M.; Fodor, S. P. A.; Gentalen, E.; Ngo, N. *J. Am. Chem. Soc.* **1997**, *119*, 5081-5090.
16. Cadet, J.; Vigny, P. In *Bioorganic Photochemistry*; Morrison, H, Ed.; Wiley; New York, 1990, Vol. 1, pp 1-272.
17. Blanchard, A. P.; Kaiser, R. J.; Hood, L. E. *Biosensors and Bioelectronics* **1996**, *11*, 687-690.

Chapter 7

Low-Density Array Optical Chip Technology for the Detection of Biomolecules

Bernard H. Schneider[1], Frederick D. Quinn[2], and David A. Shafer[3]

[1]Photonic Sensor, 2930 Amwiler Court, Suite A, Atlanta, GA 30360
[2]Tuberculosis/Mycobacteriology Branch, Centers for Disease Control and Prevention, 1600 Clifton Road, Atlanta, GA 30333
[3]Emory University, School of Medicine, 1380 South Oxford Road, Atlanta, GA 30333

Optical chips based on interferometric detection of biomolecular binding interactions provide a basis for low-density array sensors for protein and nucleic acid analysis. The refractometric sensing principle of this device allows real-time detection for analytes in the picomolar concentration range, while the use of antibody conjugates of high refractive index nanoparticles results in additional detection sensitivity enhancements of several orders of magnitude. Assay formats have been adapted to the different requirements of bioassays, including rapid, one-step immunoassays in serum and whole blood, and the discrimination of mycobacterial RNA samples without target amplification.

Refractometric optical sensors form a distinct class of sensing device, that exploit the effect of refractive index (optical density) on the propagation of guided lightwaves. The key interest in these devices stems from the universal nature of the sensing principle, and that it can be used to monitor biomolecular interactions without recourse to secondary means of signal generation such as fluorescence or chemiluminescence. When a single-mode, coherent lightwave is coupled into a waveguiding layer an evanescent field is generated during the propagation of the lightwave. This evanescent field extends on the order of one wavelength of light into the sample medium (and supporting substrate), with an exponentially decreasing intensity. This results in the majority of the field strength existing at the waveguide/solution interface. Biomolecules binding or adsorbing to this interface result in a refractive index increase due to a concentration effect over that of the sample itself. This change in refractive index affects both the angle at which light couples into the waveguide, and the speed of the lightwave propagating through the waveguiding layer.

These two effects of biomolecular concentration at the sensor/solution interface are exploited in different types of refractometric sensor, which fall typically into two groups: resonance devices, such as surface plasmon resonance (SPR) (*1-5*), the resonant mirror (*6,7*) and grating couplers (*8-10*); and interferometers (*11-19*). In resonance devices the incident angle at which the coherent light source couples into the waveguiding layer (or thin metal film in the case of SPR) is measured. Interferometers measure the change in the speed of the confined surface lightwave as a result of biomolecular binding relative to a second lightwave (or separate portion of the same lightwave) that acts as a reference. The output from an interferometer is therefore a differential signal.

Differences between these devices reside in their refractive index sensitivity, their technical implementation and their ability to null out non-specific interference. In all these areas interferometers have some advantage. The refractive index sensitivity of interferometers is $10^{-6} - 10^{-7}$, allowing the detection of pg/mm^2 levels of surface binding, typically an order of magnitude better than that of SPR and grating coupler sensors, and two orders of magnitude improvement over the resonant mirror device. This enhanced sensitivity is mainly due to the greater interaction length of interferometers, which can be 1 cm or more. Further improvement in detection sensitivity could be achieved by using techniques such as active signal processing to improve signal:noise resolution. For example, detection of pg/mm^2 levels of binding has been demonstrated for the BIAcore 2000 SPR instrument by lowering the noise level an order of magnitude over previous models (*20*). However, for analyte detection in clinical specimen matrices, background non-specific binding will define the detection limit rather than the refractive index sensitivity under ideal conditions. In resonance devices and common path interferometers the ability to null out such non-specific interference can be accomplished by using a separate sensor, which lacks the specific receptor. In non-common path interferometers a physically separate reference exists that is exposed to the test sample, and which therefore allows optical referencing to take place before the output is converted into an electrical signal. This can improve signal:noise by at least a factor of two.

While all interferometers can have essentially the same refractive index sensitivity (which is dependent on factors such as path length and refractive indexes of the waveguiding and substrate layers), there are some major differences in technical implementation. The traditional Mach-Zehnder interferometer uses channel waveguides, which are typically only several microns in width. Simple and accurate coupling into these channels is technically more difficult than coupling into planar waveguide structures. The difference interferometer (*12,13*) uses polarization optics to measure phase difference between the TE and TM modes of a planar lightwave. This common path design does not allow for a separately functionalized reference, and because the evanescent field of both modes penetrate into the sample (although to differing extents), the sensitivity of this device is reduced somewhat in comparison with the Mach-Zehnder. The Hartman interferometer (*21*) combines the simplicity of a planar lightwaves and a physically separate reference, without the difficulties associated with polarization optics or coupling into channel structures.

Optical Chip Technology

The optical chip design used in this work was based on the Hartman interferometer. A simplified laboratory test chip with a single interferometer has been used for both immunoassay and nucleic acid testing (*16,17*). As shown in Figure 1 (Side View), the optical chip used in these studies comprised a 140 nm silicon nitride waveguiding layer deposited on top of a BK-7 glass substrate using plasma-enhanced chemical vapor deposition. Grating couplers etched into the substrate were used to couple a 670 nm diode laser into and out of the waveguiding layer. The two sensing regions (15 mm in length by 1 mm wide, with a 1 mm gap) were defined using silicon dioxide (SiO_2) overlayers (Figure 1 Top View). A 500 nm thick SiO_2 film was deposited onto the optical chip everywhere except in the two sensing regions, where a thickness of 40 nm was used. This process allowed for biomolecule attachment on the sensing regions via silane chemistry, but also acted to bury the evanescent field of the propagating lightwaves except in the two sensing regions.

In the laboratory test chip the output is focused with a lens onto a slit in front of a photodetector. The fully integrated optic chip uses optical elements fabricated onto the chip to perform interferometry from an array of sensing regions (Figure 2). The input-coupled light is passed through an array of interferometers. Then, the integrated TIR mirrors and Fresnel beamsplitters are used to combine the lightwave propagating through adjacent specific and

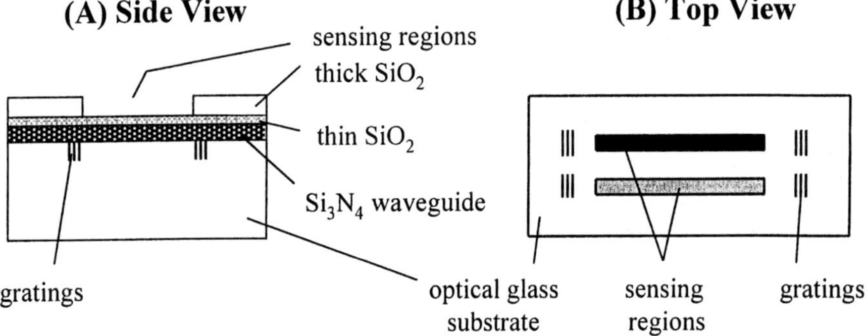

Figure 1. Schematic of laboratory test chip: (A) side view; (B) top view. The optical chip was fabricated by depositing a thin Si$_3$N$_4$ waveguiding layer on top of an optical glass substrate. A thin SiO$_2$ layer was deposited on top of the waveguiding layer in the sensing regions to provide a surface for biomolecule attachment. A thick SiO$_2$ layer was used to bury the evanescent field from the sample outside of the sensing regions.

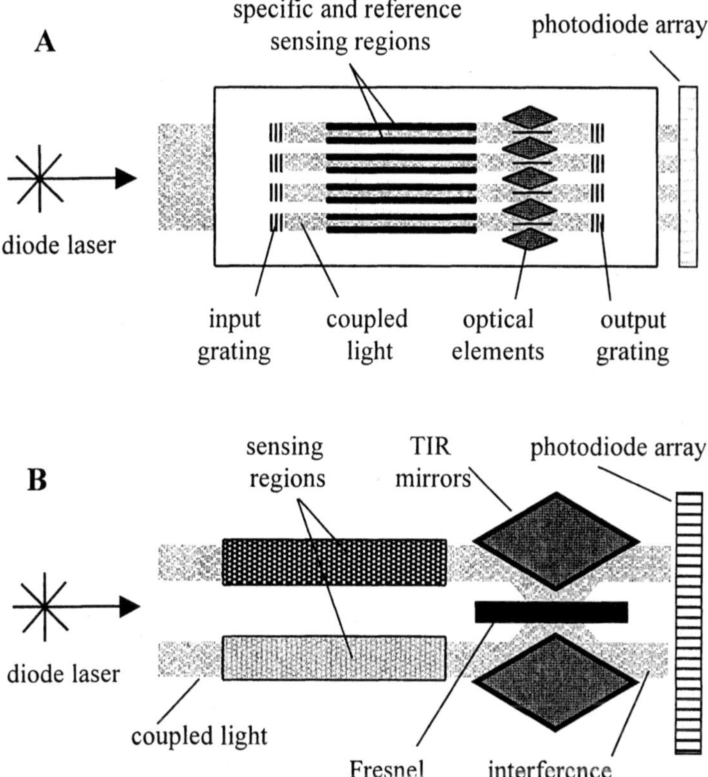

Figure 2. Integrated optic chip schematic showing: (A) multiple Hartman interferometers fabricated on a single chip; (B) expanded view of a single interferometer. TIR mirrors and Fresnel beamsplitters are used to combine adjacent portions of single coupled laser beam to form twin interferometric outputs that can be imaged directly onto a simple photodiode array positioned underneath the output gratings.

reference regions (Figure 2b) to create standing interference waves on the chip, which can then be imaged directly onto a photodiode array. The integrated format not only allows for multiple analytes to be addressed on the same chip, but also simplifies the instrument requirements to a diode laser and a low-cost photodiode detector array.

Experimental

Functionalization of Optical Chips with Biomolecular Receptors

Photolithographic patterning was used to allow specific and non-specific receptors to be immobilized on specific and reference sensing regions on the same optical chip. These procedures have been reported in detail elsewhere (*16,17*). The optical chips were cleaned with hot chromic-sulfuric acids. This cleaning solution is hazardous, and should be carried out in a fume hood with the container covered. The optical chips were removed from the acid bath, and rinsed with deionized water, only after the acid had cooled back down to room temperature. The cleaned optical chips were patterned with a positive photoresist. With photoresist remaining on the sensing regions the optical chip was silanized with a 3% solution of (tridecafluoro-1,1,2,2-tetrahydro-octyl)-1-dimethylchlorosilane in hexane. This step resulted in a hydrophobic perfluorosilane film surrounding the two sensing regions. After removing the photoresist with acetone the sensing regions were silanized with a 3% solution of 3-glycidoxypropyldimethylethoxysilane (GOPS) under mildly acidic conditions. Oxidation of the GOPS film to terminal aldehyde groups was carried out using 0.5% sodium periodate in 80% glacial acetic acid. All these steps were carried out in a fume hood.

Reductive amination chemistry has been used to couple both proteins and DNA oligonucleotides onto the activated GOPS films. Sodium cyanoborohydride was used to reduce the Schiff bases formed between amine groups on the biomolecules with the aldehyde groups on the surface to stable amide bonds. Care should be taken in handling this material, which is both very hygroscopic and toxic. After immobilization the biofilms were treated with a stabilizing solution made from casein, glucose and inositol (GIC solution), dried in an incubator at 37°C, then stored in a desiccator until tested. The use of the GIC stabilizing solution was not required with oligonucleotide probes.

Preparation of Nanoparticle Conjugates

Monoclonal antibodies (Mabs) were conjugated to different nanoparticles as reported previously (*17*). Conjugation with 15 nm colloidal gold was carried out by passive adsorption, at approximately 1 OD and pH 7. The gold nanoparticles were coated with a final Mab concentration of 20 μg/mL with gentle vortexing for 20 seconds. Blocking with 0.6% bovine serum albumin (BSA) was then carried out by vortexing gently for 60 seconds (final concentration of BSA was achieved by use of a 20% stock solution). The preparation was centrifuged at 12,000 rpm for 10 minutes, or until a soft pellet was formed with no aggregates visible. Approximately 90% of the supernatant was removed and the conjugate was resuspended in 0.6% BSA and stored at 4°C.

Test System and Assay Procedures

Measurements with dry-stored optical chips were carried out using a computer-controlled flow system (Figure 3). A methacrylate flow cell was sealed against the functionalized optical chip surface using a double-sided adhesive gasket with the center cut out to provide a thin-layer volume for the sample to flow across the two sensing regions. The thickness of this gap was approximately 100 μm. The flow cell-optical chip assembly was mounted in a slot in the test platform and held in place using a screw-down bracket. The test system was comprised of 3 syringe pumps routed through a Smart Valve (all Cavro, Sunnyvale, CA) to the test cell, then to waste. Both data acquisition and pump operation was controlled by a Macintosh® computer running LabView® software (National Instruments, Austin TX).

In all experiments background solution was flushed through the flow cell to rehydrate the chip surface, followed by fine adjustment of laser coupling to maximize signal modulation. This background solution was the same as the sample solution, but without any analyte or nanoparticle conjugate. Data acquisition was started, followed immediately by pumping the sample solution though the system. A constant flow rate of approximately 60 μl/min was used. In assay procedures using signal amplification the nanoparticle conjugate was either added to the analyte solution (1-step assay method) or pumped through the system in a second step after the analyte solution (2-step method). In some experiments the response to the sample solution was measured after the background drift had been allowed to settle out by flowing background solution through the cell until a constant rate of signal change was observed.

Results and Discussion

Real-Time Immunoassay and Nucleic Acid Analysis

The ability of refractometric sensors to measure biomolecular binding in real-time, without the need to separate the analyte from interferents, allows considerable flexibility in assay design. The sensor provides an inherently

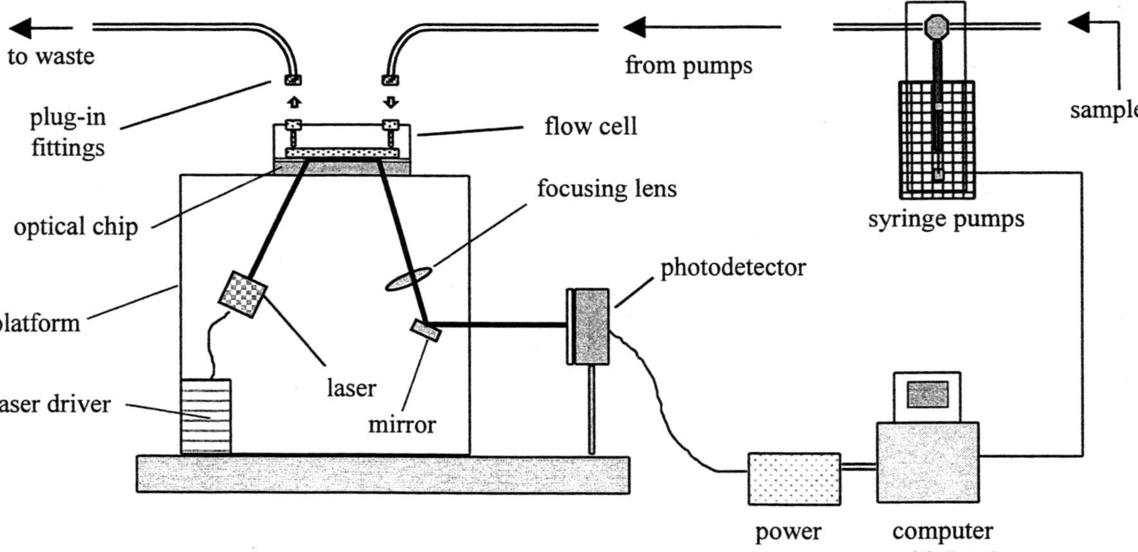

Figure 3. Schematic view of test system. An optical chip with attached flow cell was mounted in the test platform. A diode laser was coupled into the optical chip from below, and the outcoupled light was focused with a lens onto the photodetector. A computer was used for both data acquisition and to operate the syringe pumps that controlled sample flow.

quantitative output that is a function of the time interval over which the signal is integrated. Furthermore, there is no fundamental requirement for wash steps.

For large proteins at picomolar or higher concentrations ($\geq$ 1 ng/mL) direct, unamplified detection and a noncompetitive immunoassay format can be used, except where very high levels of non-specific binding (NSB) may obscure the binding signal. For low molecular weight analytes a competitive assay format can be employed to achieve close to the same levels of sensitivity (5,20,22,23). In either case, to achieve lower detection limits signal amplification using nanoparticles can be employed (17,24-26). This approach also has the advantage of counteracting the differences in size of analyte, since the response of the nanoparticle binding to the surface of the optical chip is orders of magnitude greater than that of the analyte itself. Both competitive and noncompetitive assay formats can be used. An additional degree of versatility is realized by choosing either 1-step or 2-step formats (see the following section; Signal Amplification using Nanoparticles).

Typical results for the real-time binding of an analyte to its complementary receptor immobilized on the optical chip surface are shown in Figure 4, using a model system for the detection of human chorionic gonadotropin (hCG). The measurements shown in Figure 4 were obtained in buffer solution comprising 0.2% bovine serum albumin in 20 mM phosphate-buffered saline (PBS-BSA solution). The increase in phase output from the sensor can be measured as the antigen binds to the complimentary antibody on the chip, and this binding curve changes predictably as the concentration of antigen increases (Figure 4A). Quantitative dose-response curves were obtained for both initial binding rate and equilibrium measurements (Figure 4B). The equilibrium curve demonstrates that the chip behaved like a solid phase immunoassay, with a linear slope at lower concentrations equal to unity (for the double-logarithm plot) and a saturation region at higher levels of antigen (27). When the initial binding rate is measured instead, this linear range is extended to higher concentrations, since saturation of the surface is not occurring. While Figure 4 shows an example of antibody-antigen binding, the same behavior can be observed for other biomolecular systems, such as the hybridization of a DNA oligonucleotide target to its complimentary oligonucleotide probe immobilized on the optical chip surface (16).

Signal Amplification Using Nanoparticles

Biomolecule detection at sub-picomolar concentrations with refractometric sensors generally requires the use of some form of signal enhancement, particularly from clinical specimens matrices such as serum and whole blood. This can be achieved easily by the use of high refractive index nanoparticles to provide signal amplification of biomolecular binding. This approach uses two

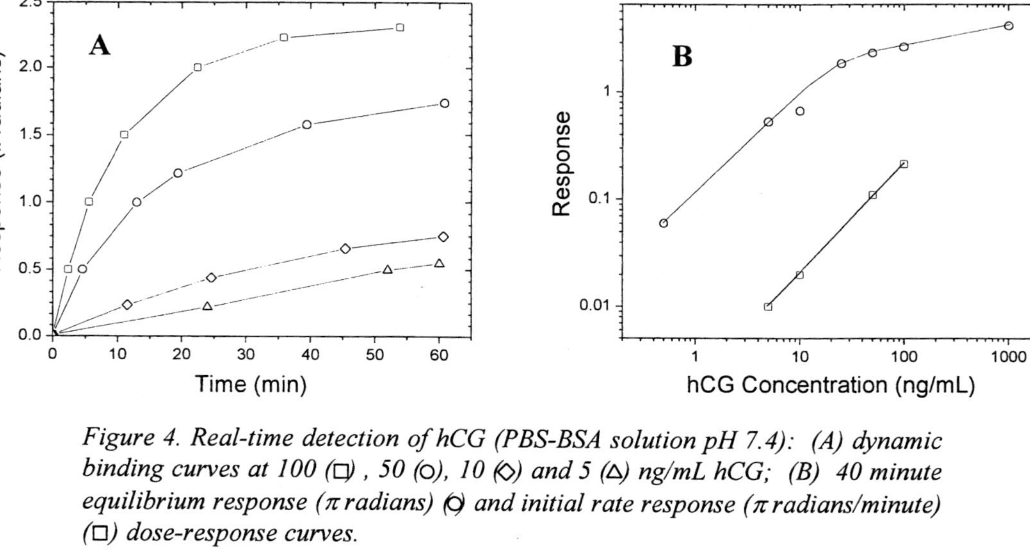

Figure 4. Real-time detection of hCG (PBS-BSA solution pH 7.4): (A) dynamic binding curves at 100 (□), 50 (○), 10 (◇) and 5 (△) ng/mL hCG; (B) 40 minute equilibrium response (π radians) (○) and initial rate response (π radians/minute) (□) dose-response curves.

receptors directed against the target of interest. The primary receptor is immobilized on the optical chip, while the secondary receptor is conjugated to the nanoparticle. Figure 5 shows the use of a 15 nm colloidal gold conjugate of an anti-hCG monoclonal antibody to improve the sensitivity of the hCG immunoassay shown in Figure 4. As indicated in the schematic, a 2-step format was used, with the first step being hCG binding to the chip (Figure 5, curve A) and the second step the measurement of the binding of the gold conjugate to the captured antigen (Figure 5, curve B). It is evident that the gold conjugate resulted in a detection level improvement of at least two orders of magnitude. In comparison tests, antibody conjugates using different types of nanoparticles, such as colloidal gold, latex and titanium dioxide, were examined for use in the hCG immunoassay. Compared to unamplified detection, 5 nm gold resulted in an enhancement factor of 22-fold, while 15 nm gold gave a 220-fold amplification, an order of magnitude better (*16*). Other nanoparticles, such as 30 nm gold, 35 nm titanium dioxide and 100 nm latex, produced enhancement factors of 290-, 228- and 250-fold respectively, on the same level of enhancement as the 15 nm gold. Subsequently, the 15 nm gold conjugate was tested more extensively in buffer, human serum (*16*) and human whole blood (*26*).

The great similarity in the observed levels of signal enhancement of all the nanoparticles tested (expect the 5 nm gold) requires some comment, and further investigation. It is expected that differences in signal enhancement would be observed when using nanoparticles of different sizes and refractive indexes. Thus, although the 100 nm latex nanoparticles have a volume more than 23 times that of the 35 nm titanium dioxide particles, the refractive index of titanium dioxide (2.6) is significantly higher than that for latex (1.58). In contrast, the refractive index of gold is too high to measure because it is non-transparent at the laser wavelength used (670 nm). Another complicating factor in trying to predict the amplified refractive index responses with different nanoparticles is light scattering and absorption losses from the nanoparticles. These effects may play a more important role in determining the signals observed than the nominal refractive index, and could explain why the responses with the 30 nm gold particles were not much larger than those measured with the 15 nm gold.

Signal amplification using nanoparticles is also applicable to nucleic acid testing. Current work is aimed at the detection and discrimination of mycobacterium species, based on the measurement of 16S ribosomal RNA without using target amplification techniques. A conjugate of 15 nm colloidal gold with an anti-biotin monoclonal antibody has been used to improve detection sensitivity for both synthetic DNA oligonucleotides and bacterial RNA samples. The basic assay method is shown schematically in Figure 6. Synthetic target oligonucleotides (41 bases) having a base sequence taken from *Mycobacterium tuberculosis* (MTB) 16S rRNA were hybridized to complementary probes on the

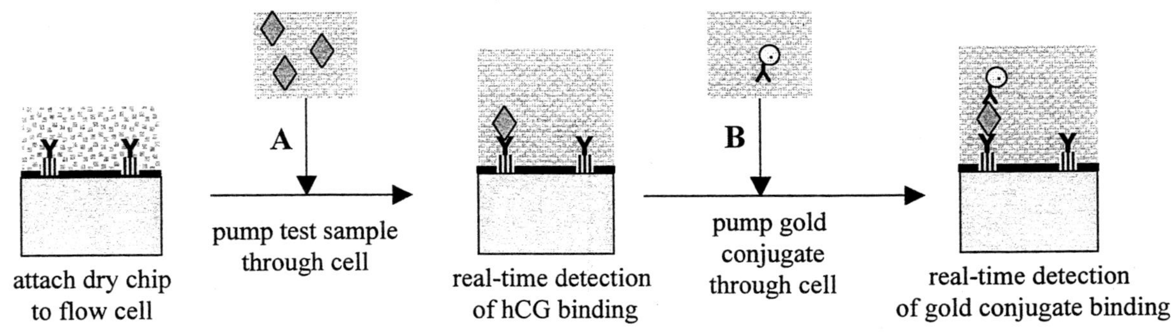
attach dry chip
to flow cell
A
pump test sample
through cell
real-time detection
of hCG binding
B
pump gold
conjugate
through cell
real-time detection
of gold conjugate binding

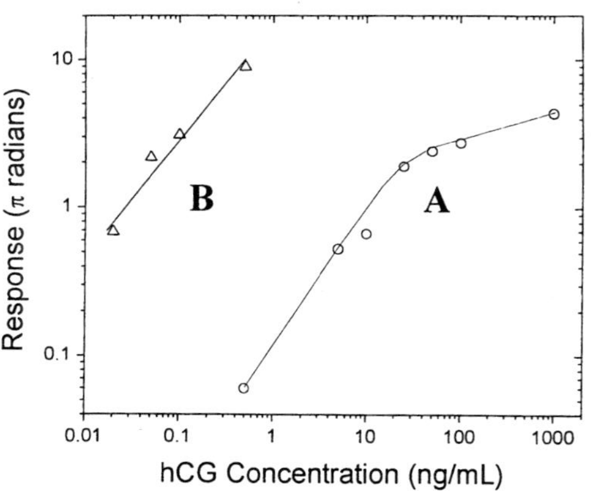

Figure 5. hCG detection in buffer comparing (A) unamplified detection; (B) two-step signal amplified assay: (○) 40 minute equilbrium measurements for unamplified detection; (△) 2-step signal-amplified assay using 15 nm gold conjugate with a total assay time of 65 minutes (50 minute hCG binding step followed by a 15 minute gold conjugate binding step). Top schematic illustrates the binding of hCG to the capture antibody film on the chip surface, followed by detection of the gold conjugate. (From ref. 26, with permission from Elsevier Science.)

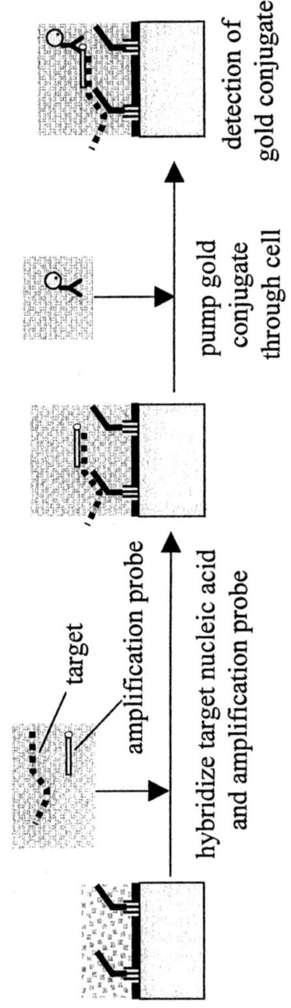

target
amplification probe
hybridize target nucleic acid and amplification probe
pump gold conjugate through cell
detection of gold conjugate

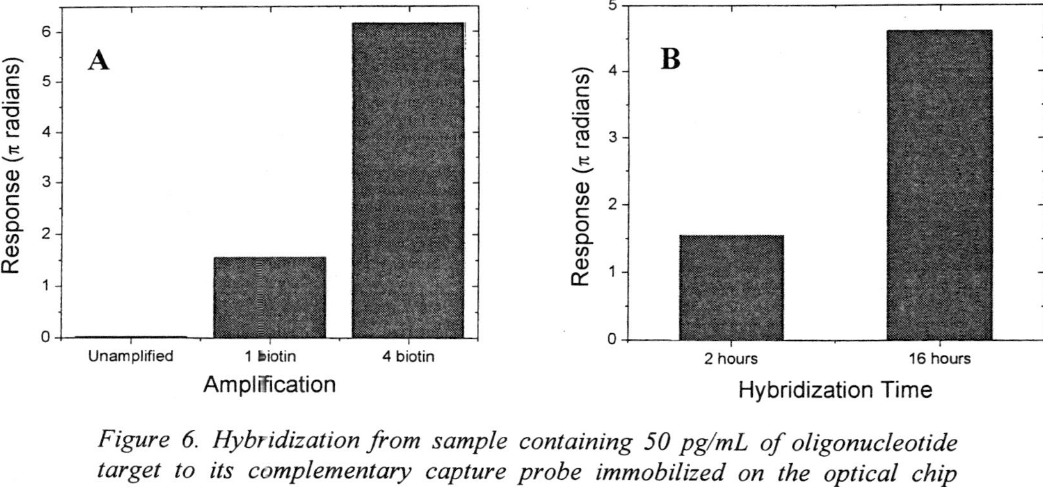

Figure 6. Hybridization from sample containing 50 pg/mL of oligonucleotide target to its complementary capture probe immobilized on the optical chip surface: (A) 2 hr hybridization measurements comparing unamplified versus signal-amplified detection with an amplification probe containing either 1 or 4 biotins; (B) comparison of signal-amplified response after different hybridization times of 2 or 16 hours using an amplification probe containing 1 biotin. Top schematic illustrates assay procedures. The amplification probe is hybridized to the target simultaneous with the hybridization of the target to the capture probe. After an optional wash step to improve hybridization stringency, the amount of target hybridized is then detected by the response from an anti-biotin:gold conjugate.

optical chip surface in the presence of a second oligonucleotide (amplification probe) that was labeled with biotin. This amplification probe was designed to be complementary to the target immediately adjacent to the capture region. In the second step the binding of the anti-biotin:gold conjugate to the amplification probe was measured. A stringent wash in between these steps allowed improved discrimination between the MTB-sequence target and a closely related *Mycobacterium avium*-sequence (MAV) target (4 base difference). Figure 6A shows graphically the level of signal amplification possible. Using an amplification probe with a single biotin provided an enhancement of about 150-fold over unamplified detection, while using an amplification probe that incorporated 4 biotin molecules gave a signal level about 4 times higher. Further gains would be expected using amplification probes with even higher numbers of incorporated biotins. The results in Figure 6A were obtained for 2 hour hybridizations at room temperature. When this hybridization time was increased to a 16 hour step (overnight hybridization), a roughly three-fold increase in response was observed (Figure 6B).

It should be noted that refractive index measurements facilitate the detection of multiple analytes, even when 1-step or 2-step signal amplification schemes are employed. This technical simplicity results from the localized nature of refractive index change, which is measurable only where surface binding takes place. In contrast, the spherical radiation of light emission resulting from fluorescence or chemiluminescence results in cross-talk problems that make spatial analysis of multiplex arrays difficult, and which typically require complex instrumentation and signal processing to deconvolute.

Immunoassay Measurements in Whole Blood

The ability to measure biomolecular binding continuously in real-time allows the amplified assay to be performed in a 1-step format, an option not possible with techniques such as fluorescence or chemiluminescence. In this format (Figure 7 schematic) the gold conjugate is added to the sample, where it binds to the hCG in solution. The binding of the formed hCG-conjugate complex to the capture antibody on the optical chip can then be measured as it happens. There was found to be no appreciable difference in response for the 1-step and 2-step methods when the total assay time was kept constant (*26*). Dose-response curves for hCG detection in serum and whole blood have been carried out using this 1-step assay format, and the results of the whole blood measurements are shown in Figure 7. Figure 7A shows the response for 1 ng/mL, 0.5 ng/mL and 0 ng/mL (control) samples as a function of the total assay time, while the response as a function of hCG concentration for the 10 minute assay is shown in Figure 7B. Discrimination of 0.5 ng/mL hCG from the control was obtained after only 10 minutes, although experimental variance (in large part due to the high background observed in whole blood) would have prevented the resolution of intermediate concentrations of analyte. At longer assay times the signal: noise ratio improved.

Detection and Discrimination of Bacterial Ribosomal RNA

The same approach has been used to detect and discriminate MTB 16S rRNA from MAV and *E. coli* RNA (Figure 8), using an amplification probe with a single biotin, and overnight hybridization step. Excellent discrimination was observed, with the response for MAV RNA at about 3% of the specific MTB response, while the *E. coli* response was an order of magnitude lower than the MAV response. The observed response for MTB 16S rRNA at 300 ng/mL total bacterial RNA is at least 2 orders of magnitude above the optical chip resolution limit. Also, amplification probes with greater numbers of incorporated biotins would allow even greater improvements in the detection limit of the assay.

Conclusions

Results to date have confirmed that the Hartman interferometer provides an effective platform on which refractometric optical chip sensors can be developed for clinical diagnostic and other biological testing applications. Rapid and sensitive immunoassay measurements have been made in clinical specimens such as human serum and human whole blood. Several features of the sensor allow for simplified assay procedures: there is no fundamental requirement for wash or separation steps; the optical reference compensates for background noise; and the localized nature of refractive index measurement makes multiplex detection using signal amplification simpler than the corresponding fluorescence analysis. The optical chip has been used to sense a range of biomolecules, including proteins (hCG, IgG immunoglobulins, hepatitis B surface antigen), pathogens (*Salmonella typhimurium*, influenza A virus) and nucleic acids (synthetic oligonucleotides, mycobacterium RNA). Current detection limits are already in the picogram/mL range, and further improvement is expected through assay optimization, enhanced signal amplification and active signal processing.

Acknowledgments

The work described was supported in part by grants 1 R43 AI34793-01A1 and 1 R43 AI39924-01 from the National Institutes of Health. The contents of this paper are solely the responsibility of the authors and do not necessarily represent the official views of the National Institutes of Health.

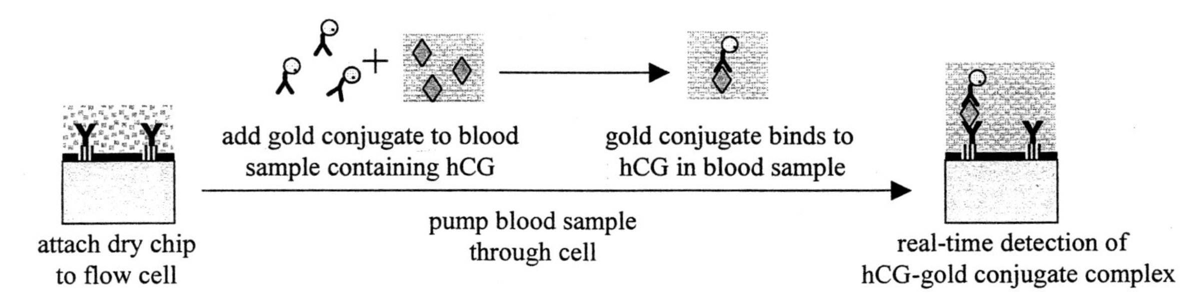

add gold conjugate to blood
sample containing hCG
gold conjugate binds to
hCG in blood sample
attach dry chip
to flow cell
pump blood sample
through cell
real-time detection of
hCG-gold conjugate complex

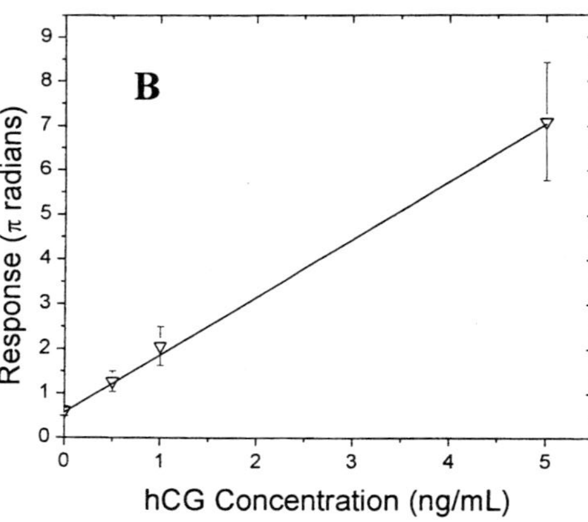
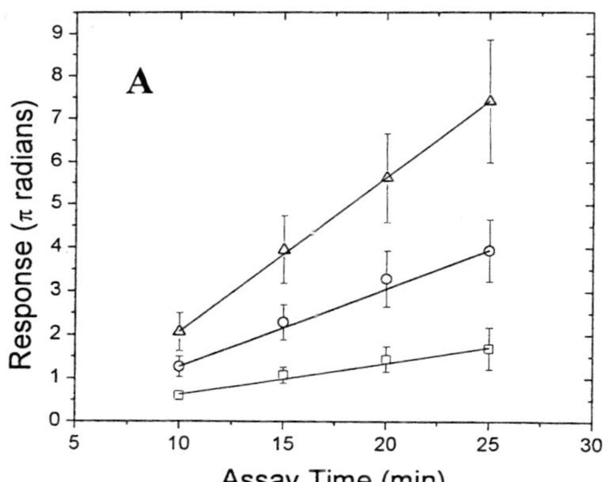

Figure 7. hCG detection in whole human blood using a 1-step signal-amplified assay: (A) the binding response measured as a function of assay time for 0 (□), 0.5 (○) and 1 ng/mL (△) of hCG; (B) the 10 minute dose-response curve plotted as a function of hCG concentration. Additional data obtained at 5 ng/mL is shown. Top schematic shows the assay procedure. A 15 nm gold conjugate of an anti-hCG Mab is added to the blood sample immediately prior to testing. After rehydration of the optical chip the sample is pumped through the test cell and the binding of the hCG:gold conjugate complex is measured in real-time. (From ref. 26, with permission from Elsevier Science.)

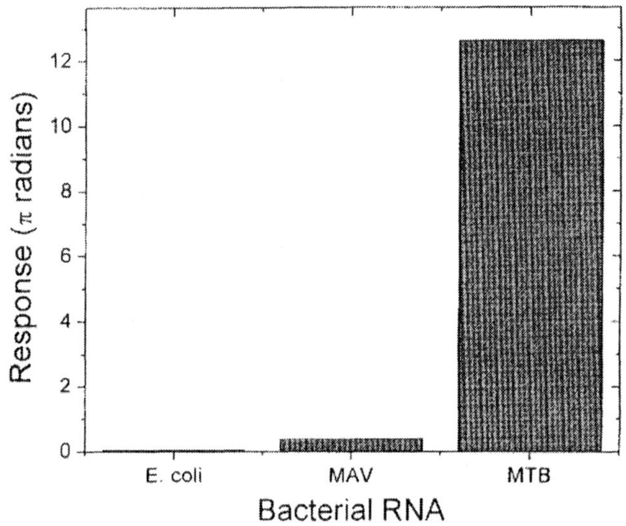

Figure 8. Detection and discrimination of bacterial 16S ribosomal RNA using a 2-step signal-amplified assay with an amplification probe containing 1 biotin. The plot shows the response for the anti-biotin:gold conjugate after overnight hybridization at a concentration of 300 ng/mL total RNA extracted from M. tuberculosis (MTB), M. avium (MAV) and E. coli cultures. The assay method is the same as that illustrated in Figure 6. A stringent wash step was used after the hybridization step to improve discriminaton between the different bacterial samples.

References

1. Lundström, I. *Biosensors Bioelectron.* **1994**, *9*, 725.
2. Lyon, L.A.; Musick, M.D.; Natan, M.J. *Anal. Chem.* **1998**, *70*, 5177.
3. O'Meara, D.; Nilsson, P.; Nygren, P.-A.; Uhlén, M.; Lundeberg, J. *Anal. Biochem.* **1998**, *255*, 195.
4. Woodbury, R.G.; Wendin, C.; Furlong, C.E. *Biosensors Bioelectron.* **1998**, *13*, 1117.
5. Harris, R.D.; Luff, B.J.; Wilkinson, J.S.; Piehler, J.; Brecht, A.; Gauglitz, G.; Abuknesha, R.A. *Biosensors Bioelectron.* **1999**, *14*, 377.
6. Cush, R.; Cronin, J.M.; Stewart, W.J.; Maule, C.H.; Molloy, J.; Goddard, N.J. *Biosensors Bioelectron.* **1993**, *8*, 347.
7. Buckle, P.E.; Davies, R.J.; Kinning, T.; Yeung, D.; Edwards, P.R.; Pollard-Knight, D.; Lowe, C.R. *Biosensors Bioelectron.* **1993**, *8*, 355.
8. Tiefenthaler, K.; Lukosz, W. *J. Opt. Soc. Am. B* **1989**, *6*, 209.
9. Fattinger, Ch.; Mangold, C.; Gale, M.T.; Schütz, H. *Opt. Eng.* **1995**, *34*, 2744.
10. Clerc, D.; Lukosz, W. *Biosensors Bioelectron.* **1997**, *12*, 185.
11. Heideman, R.G.; Kooyman, R.P.H.; Greve, J. *Sensors Actuat. B* **1993**, *10*, 209.
12. Fattinger, Ch.; Koller, H.; Schlatter, D.; Wehrli, P. *Biosensors Bioelectron.* **1993**, *8*, 99.
13. Schlatter, D.; Barner, R.; Fattinger, Ch.; Huber, H.; Hübscher, J.; Hurst, J.; Koller, H.; Mangold, C.; Müller, F. *Biosensors Bioelectron.* **1993**, *8*, 109.
14. Brosinger, F.; Freimuth, H.; Lacher, M.; Ehrfeld, W.; Gedig, E.; Katerkamp, A.; Spener, F.; Cammann, K. *Sensors Actuat. B* **1997**, *44*, 350.
15. Schipper, E.F.; Brugman, A.M.; Dominguez, C.; Lechuga, L.M.; Kooyman, R.P.H.; Greve, J. *Sensors Actuat. B* **1997**, *40*, 147.
16. Schneider, B.H.; Edwards, J.G.; Hartman, N.F. *Clin. Chem.* **1997**, *43*, 1757.
17. Schneider, B.H.; Dickinson, E.L.; Vach, M.D.; Hoijer, J.V.; Howard, L.V. *Biosensors Bioelectron.* **2000**, *15*, 13.
18. Luff, B.J.; Wilkinson, J.S.; Piehler, J.; Hollenbach, U.; Ingenhoff, J.; Fabricus, N. *J. Lightwave Technol.* **1998**, *16*, 583.
19. Weisser, M.; Tovar, G.; Mittler-Neher, S.; Knoll, W.; Brosinger, F.; Freimuth, H.; Lacher, M.; Ehrfeld, W. *Biosensors Bioelectron.* **1999**, *14*, 405.
20. Karlsson, R.; Ståhlberg, R. *Anal. Biochem.* **1995**, *228*, 274.

21. Hartman, N.F. *US Patent 5,633,561* **1997**.
22. Bier, F.F.; Schmid, R.D. *Biosensors Bioelectron.* **1994**, *9*, 125.
23. Schipper, E.F.; Bergevoet, A.J.H.; Kooyman, R.P.H.; Greve, J. *Anal. Chim. Acta* **1997**, *341*, 171.
24. Watts, H.J.; Lowe, C.R.; Pollard-Knight, D.V. *Anal. Chem.* **1994**, *66*, 2465.
25. Kubitschko, S.; Spinke, J.; Brückner, T.; Pohl, S.; Oranth, N. *Anal. Biochem.* **1997**, *253*, 112.
26. Schneider, B.H.; Dickinson, E.L.; Vach, M.D.; Hoijer, J.V.; Howard, L.V. *Biosensors Bioelectron.* **2000**, *15*, 597.
27. Peterman, J.H. In *Immunochemstry of Solid-Phase Immunoassay*; Butler, J.E. Ed.; CRC Press: Boca Raton, **1991**; Chapter 3.

Chapter 8

Randomly-Ordered High-Density Fiber Optic Microsensor Array Sensors

Jason R. Epstein, Shannon E. Stitzel, and David R. Walt*

The Max Tishler Laboratory for Organic Chemistry, Department of Chemistry, Tufts University, 62 Talbot Avenue, Medford, MA 02155

By combining fiber optic arrays with fluorescence-based detection techniques, high-density sensors for vapor- and solution-phase analytes were developed. These research formats have been employed for both organic vapor discrimination and DNA detection. Organic vapor discrimination entails fingerprinting odors via the sensor array's temporal response patterns, mimicking vertebrate olfaction. These microsphere-based artificial nose systems can discriminate common organic vapors or detect explosive-like compounds. The fiber optic DNA detection platform is an alternative microarray platform for high throughput genomic interrogation techniques. This DNA detection format can incorporate a variety of sensing designs, including molecular beacons, and is proficient in single nucleotide polymorphism detection. Both fiber optic detection platforms have simple fabrication protocols and take advantage of signal averaging across like sensors, providing a signal-to-noise benefit. High-density microarrays platforms also have applications in live cell, protein, and enzymatic assays.

Optical fibers consist of a central dielectric core surrounded by an exterior material, or clad, of lower refractive index. The refractive index difference between the core and clad materials allows light to be transmitted down the entire length of the fiber through total internal reflection. The ability to carry a signal over long fiber lengths allows the fiber-optic platform to be used in remote sensing applications that typically cannot be addressed by other analytical techniques. Optical fibers can be used to directly monitor changes in the analyte's optical properties. Alternatively, fibers can be converted into sensors to indirectly monitor the analyte by using an indicator immobilized on the fiber surface (1). Combining optical fibers with fluorescence-based sensing techniques enables multiple optical properties to be monitored simultaneously, including wavelength, intensity, or lifetime. Microarrays consist of thousands of individually addressable sensors, and while there are many sensor substrates, fiber-optic arrays offer numerous advantages over other array platforms.

Individual optical fibers can be fused together to form 0.5-1.0 mm diameter imaging fiber bundles, with each fiber carrying isolated signals. These imaging fiber bundles, containing from 6,000 to 50,000 fibers, are coherent arrays that transmit images over the fiber length, as illustrated in Figure 1. By associating a fluorescence-based sensor with each fiber in the bundle, an individually addressable sensor array is created that combines the benefits of imaging fibers with optical sensing techniques. Fiber-optic sensor array applications include, but are not limited to vapor sensing, ion sensing, biosensing and organic compound sensing (1). The fiber-optic sensors in this chapter employ high-density microarray sensor designs both as vapor phase and as solution-based sensors and biosensors. Vapor phase detection is employed for organic vapor determination, such as detection of explosive-like compounds. The solution-phase experiments, primarily for DNA detection, have shown proficiency in detecting single-point mutations. These fiber optic microarray sensors are currently in development for commercial distribution (2).

Fiber-Optic Array Fabrication

A fiber-optic microarray platform is created by first forming microwells on the array's distal face, in which each microwell serves as a container for an individual sensor. Polished imaging fiber tips are selectively etched using a buffered hydrofluoric acid solution to create a high-density array of microwells (3). *Caution: hydrofluoric acid is extremely caustic!* Wells are formed due to the difference in etch rates between the core material and the clad. The etch rate and well depth depend on the fiber composition, the etching solution strength, and the length of time the fiber is exposed to the solution. Typical well diameters are 3-5 μm; complementary-sized microsensors are distributed into the wells such that each well contains only one microsphere. Atomic force micrographs of the empty wells and wells containing microspheres are shown in Figure 2. Different microsensors are combined to form a stock of sensors, and then the stock is applied to the etched fiber face, either suspended in solution or as a dry slurry. The different sensor types are mixed and randomly distributed

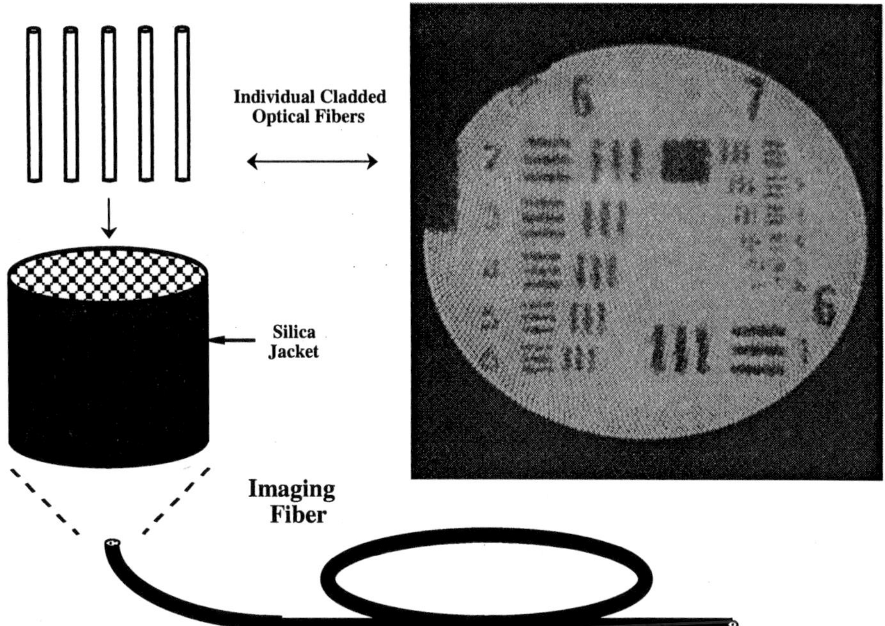

Figure 1. A coherent optical imaging fiber. An imaging fiber is a flexible fiber array comprised of thousands of individually-cladded optical fibers that are melted and drawn together such that an image can be carried from one face to the other. The top image of an Air Force resolution target acquired with a CCD camera, as viewed through a 350-µm diameter modified imaging fiber. Reprinted from ACS Symp. Ser. (1998), 690, (Polymers in Sensors), 273-289.

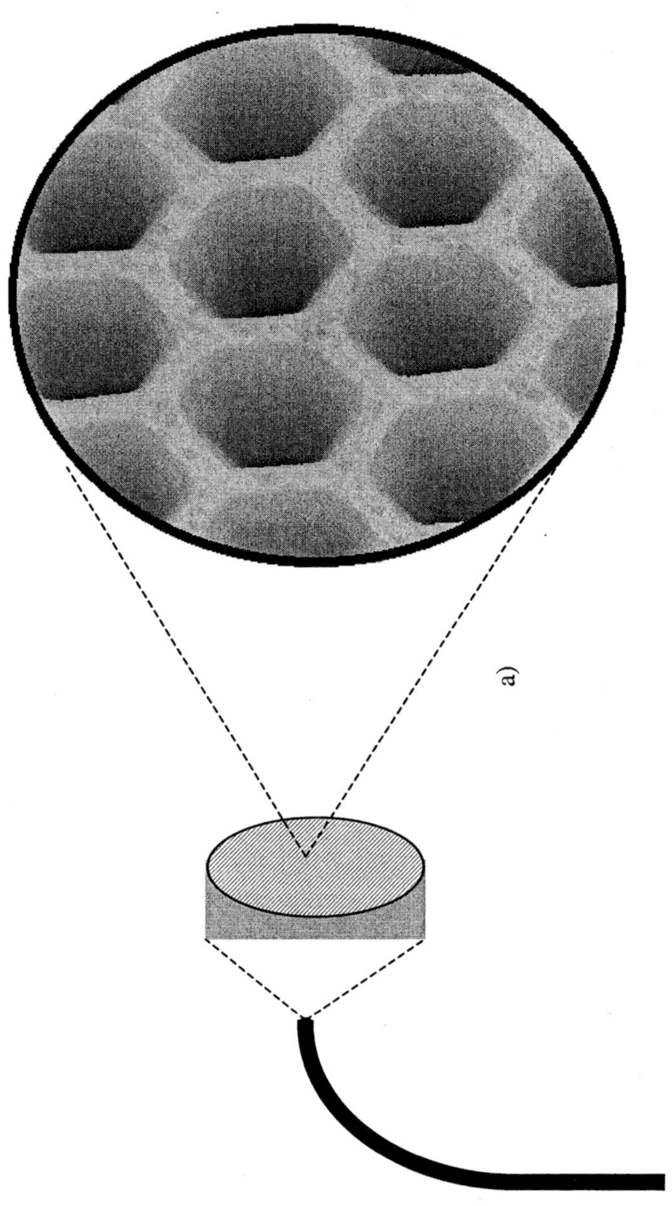

a)

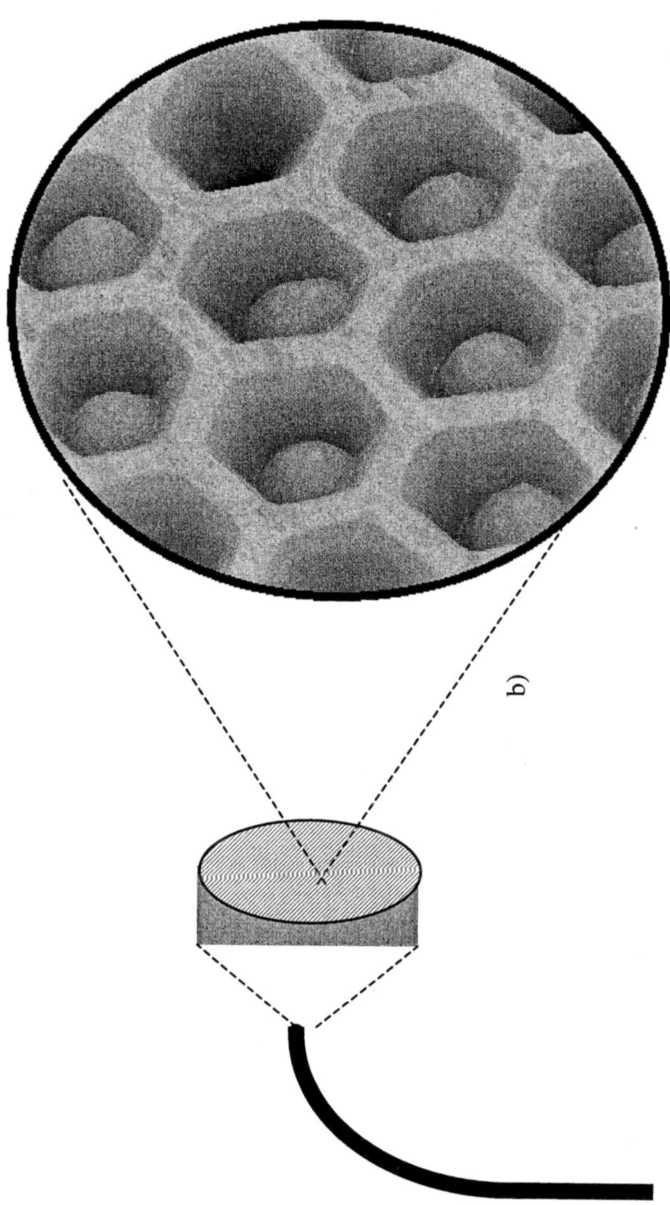

Figure 2. Images from a scanning force micrograph of a) empty wells of an etched fiber, and b) beads distributed into the wells. Well diameters are approximately 5 μm.

on the etched array with replicates of each sensor-type on every array. The extent to which each sensor-type is represented in the array is dependent on the composition of the sensor stock. The presence of multiple identical sensors provides a signal-to-noise ratio advantage, that increases by the square root of the number of sensor monitored. The random distribution process makes it necessary to positionally register the array after fabrication, and multiple decoding schemes have been developed to catalogue the position of each microsensor (4-6).

Instrumentation

All applications use a custom built epifluorescence imaging system with a xenon arc lamp excitation source. The imaging system is shown in Figure 3, and includes excitation and emission filter wheels combined with a modified microscope and a charge-coupled device (CCD) camera (Cooke Corporation, Auburn Hills, MI) or an intensified CCD (Princeton Instruments, Trenton, NJ). For fluorescence measurements, indicators on the fiber's distal tip are interrogated by attaching the fiber's proximal end to the modified epifluorescence microscope. The excitation source sends filtered light through the proximal end of the fiber, exciting either the analyte or fluorescent indicators located at the fiber's distal face. An excited fluor emits longer wavelength light, which is coupled back through the fiber, filtered and directed to the CCD detector.

Artificial Nose

The vertebrate olfactory system is capable of discriminating thousands of different odors. The ability to recognize many different scents is derived from the cross-reactive nature of the odor receptors in the olfactory epithelium. Every smell elicits responses from multiple receptors, creating a unique response pattern that encodes each odor (7,8). The brain learns by correlating the receptor response patterns associated with each smell. Since the receptors are not selective for specific odors, but are cross-reactive, thousands of possible receptor response combinations can encode individual odors. The large number of possible response pattern combinations enable the vertebrate olfactory system to discriminate a wide variety of odors, including ones that have not previously been encountered (9). Artificial noses mimic many of the principles demonstrated by vertebrate olfaction to develop widely responsive vapor detection systems (10,11). Current artificial nose systems are made with different types of sensor elements including surface acoustic wave devices

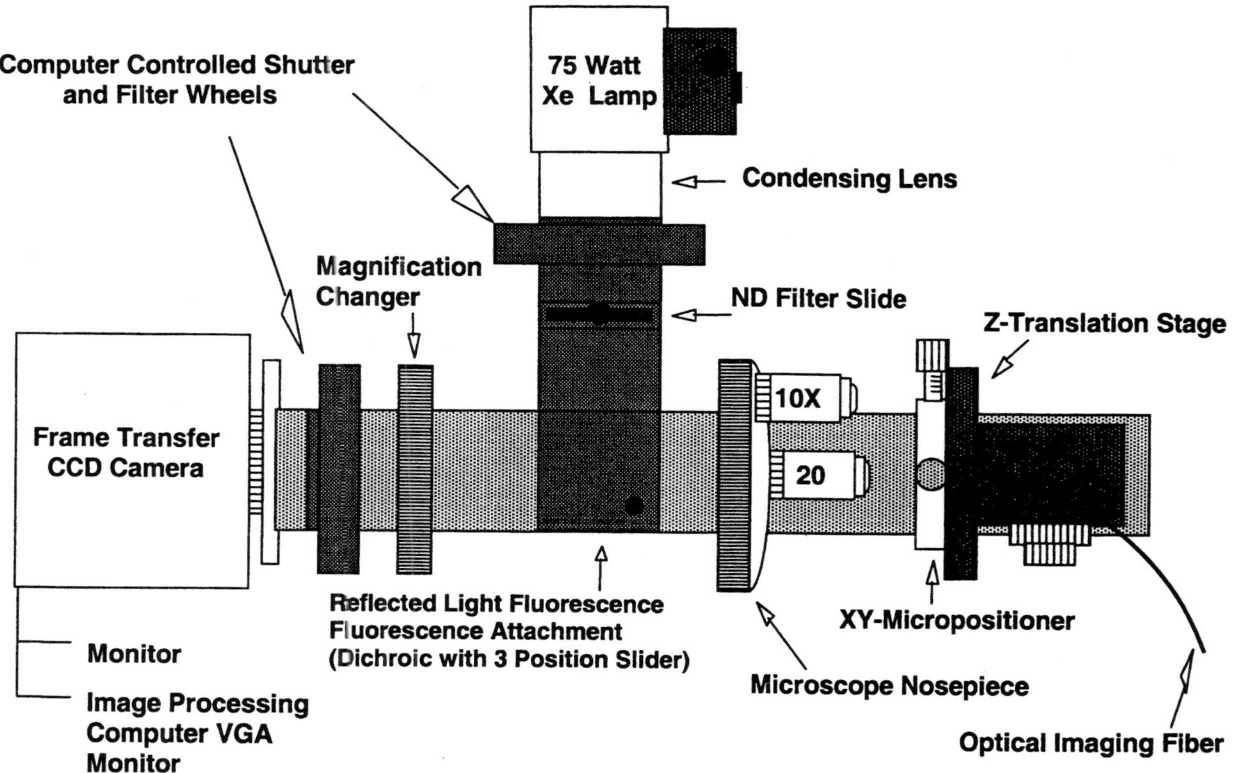

Figure 3. Instrumentation used for fluorescence measurements and imaging with optical fibers.

(12,13), conducting polymers (14,15), metal oxide sensors (16-18), carbon black-polymer composite chemiresistors (19-21), and fiber-optic based sensors (6). All of these sensor arrays reproduce the cross-reactive nature of the vertebrate olfactory receptors. The fiber-optic approach utilizes a high-density array with thousands of sensor elements, mimicking the large number of odor receptors in the vertebrate system.

The fiber-optic based artificial nose system is composed of an etched fiber-optic imaging bundle containing thousands of 3 μm diameter microsphere vapor sensors. Microsphere sensors are made from commercially available polymer (Bangs Laboratories, Fishers, IN) or silica beads (Phenomenex, Torrance, CA) and are prepared by adsorbing an environmentally sensitive fluorophore, Nile Red, to the bead surface (6). Nile Red is a solvatochromic dye (22); its fluorescence properties are affected by changes in the polarity of the dye's environment. By absorbing the dye on beads containing different surface functionalities, the various bead types exhibit different emission maxima corresponding to the polarity of the functional groups on the bead. The emission maximum and the intensity of the dye on each microsphere shifts when the array is exposed to different polarity analytes as seen in Table 1. By creating bead sensors with various functionalities, a library of cross-reactive sensors is formed, which responds to a variety of analytes.

Microsensors are fabricated in a batch process, creating reproducible stocks of approximately 1.0×10^8 beads per mg of microspheres (6). High-density sensor arrays are formed by randomly distributing a combination of sensor types into microwells on the etched tip of an optical imaging fiber, creating an array with hundreds of each sensor type (5,6). The random distribution of beads in the wells simplifies array fabrication by eliminating the need to individually align each sensor within the array. This random distribution of sensors also mimics the semi-random position of odor receptors in the vertebrate olfactory epithelium (9,23).

Fiber-optic nose arrays are monitored with the fluorescence imaging system previously discussed. Vapors are delivered to the array as a pulse via a vacuum controlled sparging apparatus (24). Each vapor pulse is 1-2 seconds long and mimics the action of sniffing. Delivering analytes as vapor pulses differs from most other artificial nose systems that rely either on equilibrium responses upon exposure to the vapor phase analyte or on single time point measurements (12,21). By monitoring the fluorescence of individual microsphere sensors before, during, and after a vapor pulse is introduced to a sensor array, temporal response patterns are recorded that are unique for each analyte, as seen in Figure 4 (25). Different sensor responses for a given analyte are due, in part, to the shift in the sensor's fluorescence with the change in environment's polarity. Response patterns are also affected by the emission wavelength monitored, the partition coefficient of the analyte, and swelling effects of the polymer beads.

Table 1. Shift in emission maxima of different sensor types in the presence vapors with different polarities.

	Air	Methanol	Ethanol	Toluene	Heptane
Bead-OH	650	644	642	652	664
Bead-CN	624	642	638	636	638
Bead-CH$_3$	618	638	638	618	618

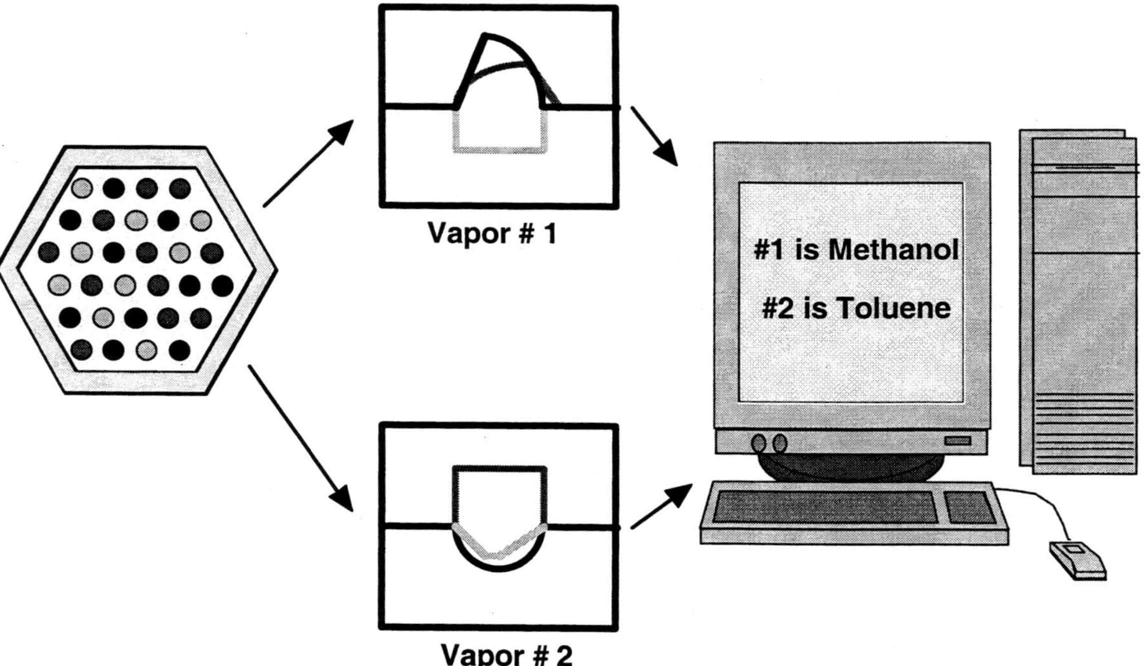

Figure 4. Diagram of the pattern recognition process with a cross-reactive vapor sensor array. Each vapor is encoded by the unique temporal response patterns elicited from the different sensor types in the array. Vapor 1 is characterized by two positive response curves and Vapor 2 is characterized by two negative response curves. The differences in the overall response patterns for each vapor are learned and recognized by a computational network.

Each bead type's temporal response patterns are used primarily to identify different analytes, but are also used as a self-encoding mechanism by which sensor types can be identified (6). As previously described, the different sensor types are randomly distributed throughout the array, and therefore the array must be registered to identify the position of the different sensor types. Registration is accomplished by exposing sensor arrays containing the individual bead types to high concentrations of acetone, ethanol or other volatile organics. By comparing responses of the randomized sensors to the known responses of the individual bead types, the randomized array is positionally registered. Once an array is registered, data are collected and analyzed using a variety of computational approaches, including principal components analysis, nonparametric statistics, neural networks and genetic algorithms. These computational networks are trained to correlate the sensor response patterns with unidentified analytes through repeated exposure of the sensor array to known analytes (See Figure 4).

Microsphere sensor arrays improve the sensing capability of the optical nose system because the sensor redundancy in the arrays allows signals from multiple sensors to be averaged. Averaging responses from multiple sensors creates smoother signals by eliminating some of the sensor-to-sensor variation as seen in Figure 5. Signal averaging also increases the array sensitivity because of the reduced percent coefficient of variation (6), which is one way that vertebrate noses are thought to gain sensitivity (11).

The optical nose has a wide range of possible applications. Initial work has applied a modified version of the fiber-optic nose system to the detection of explosives-like vapors. A portable system was designed that incorporated microsphere sensors on seven single core optical fibers instead of the high-density array. The portable system was shown to be sensitive to multiple explosive-like compounds and was used in field studies to detect buried land mines (26). The non-selective nature of the artificial nose sensors allows other odor recognition problems to be addressed. For example, we have used the nose to investigate discrimination of various volatile organic compounds including alcohols, aromatics, and common solvents (6,24,27). Optical noses could also be used as quality control devices, signaling an abnormality in an industrial process or a change in an environmental condition. The optical nose might also be used for more complex recognition problems such as monitoring multiple components from automobile emissions, or detecting disease through breath analysis. In addition to applying the artificial nose to real world problems, the artificial system is being used to model and increase our understanding of vertebrate olfaction.

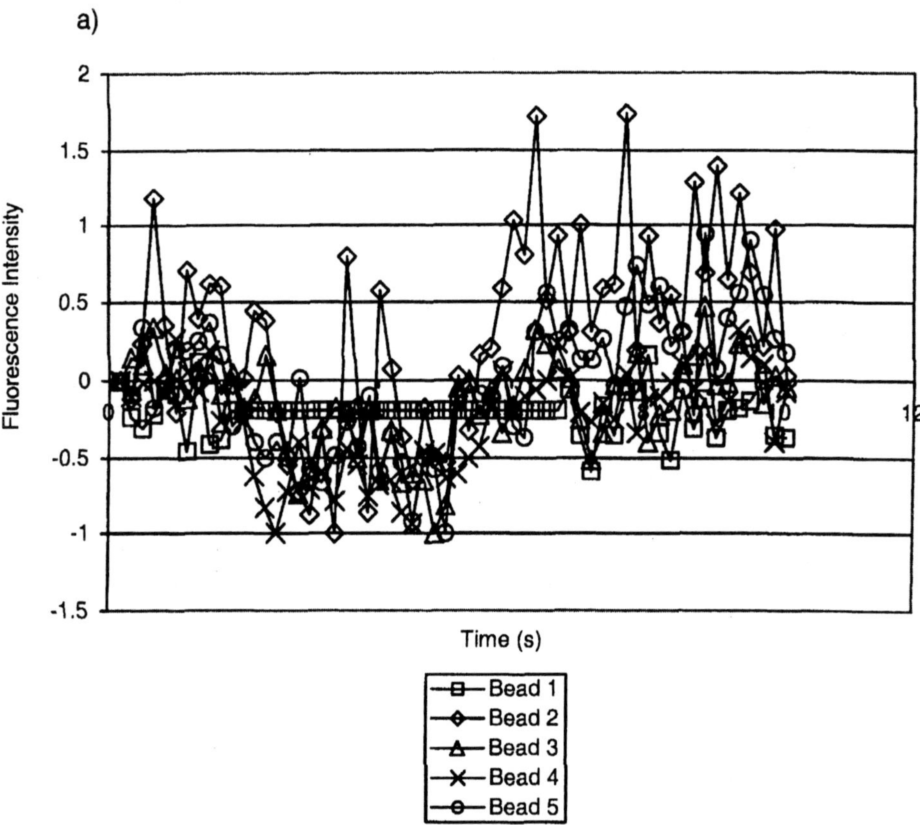

Figure 5. Comparison between individual bead responses and the sum of 100 bead responses. a) Five individual bead responses to 1.8% saturated methanol vapor. Each bead response is very noisy and the five responses are not identical. b) Summed responses of five random groups of 100 beads in response to 1.8% saturated methanol vapor. Each group of 100 beads gives a strong negative signal. This signal is independent of which 100 beads are summed as evidenced byagreement between. These figure show that there is a marked improvement in the signal-to-noise ratio by summing the signals of individual bead sensors, as well as eliminating bead-to-bead signal variation.

b)

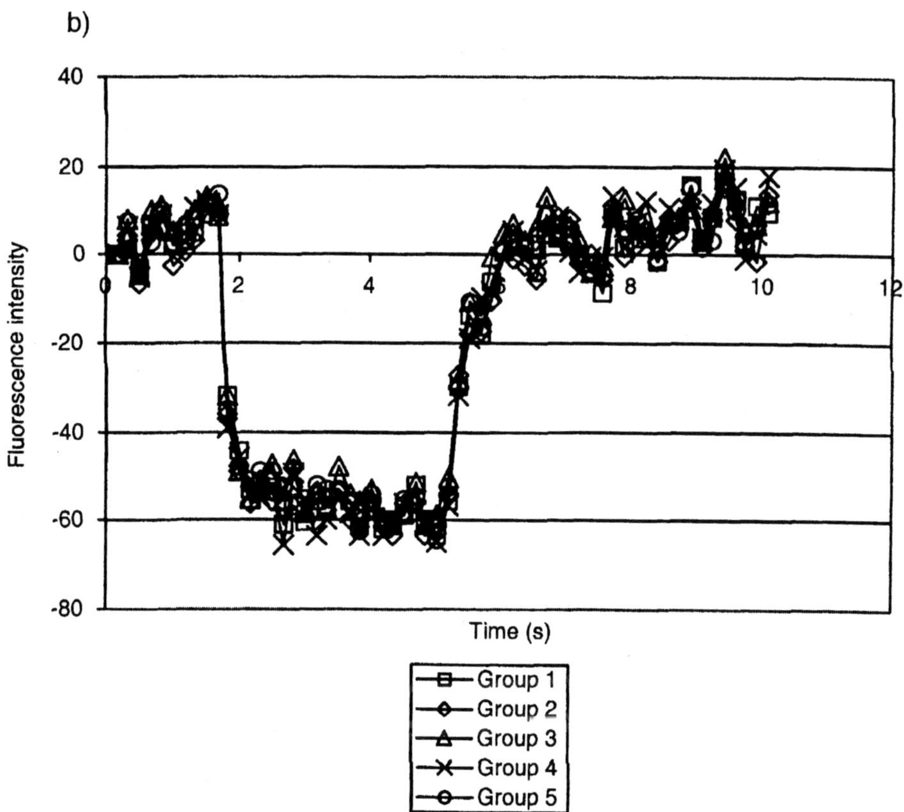

Figure 5. *Continued*

DNA Genosensors

High-density DNA microarrays have contributed significantly to biotechnology. Fiber-optic microarrays are an attractive alternative microarray platform for high throughput DNA analysis because of the high-density of individual sensing elements, along with the ability to simultaneously register multiple signals (28). Unlike other high-density array designs, fiber-optic microsensor arrays have a small overall feature size and relatively simple fabrication protocols (5). This sensor fabrication method allows new sequences of interest to be incorporated readily into the array at rather trivial expense, and new sensing elements can be added to preexisting arrays without altering the current sensing capabilities. The fiber optic arrays have detection limits of 10^{-21} mol (29) and can detect single-nucleotide polymorphisms (SNP's) (30).

Microsensor preparation

High-density fiber-optic DNA microarrays are fabricated by inserting oligonucleotide-functionalized microspheres into the etched wells of optical fibers. A large number of microsensors are made at one time; a 1 mL suspension of 3-μm beads contains up to 10^{10} beads. The microsensors are prepared by covalently immobilizing the desired probe sequence directly to the microsphere surface (4,30-32). Sensing elements are comprised of oligonucleotide sequences that exploit the fundamental principles of DNA base-pair formation via complementary-strand binding (base pairing A to T and G to C). The oligonucleotide microsensors are prepared by allowing the microspheres (Bang's Labs, Fishers, IN) containing a surface amine functionality to react with glutaraldehyde and polyethyleneimine (PEI) to amplify the number of reactive amino groups and to distance the probe from the microsphere surface (31,32). Oligonucleotide probe sequences are prepared with a C6-modifier spacer arm (Glen Research, Sterling, VA), and activated with cyanuric chloride, which allows the oligonucleotide to be coupled to the PEI (31,32). A mixture of microsensor types can be generated and combined to form a library stock of functionalized beads. Aliquots of this aqueous microsensor stock suspension are added directly to the etched face of the fiber. When excess solvent evaporates, the microspheres are randomly distributed into the microwells. The number of each oligonucleotide probe-type in the array is dependent on its relative amount in the stock suspension. The replicate microsensors allow for bead summing, which improves the signal-to-noise ratio. Since the microsensor placement occurs randomly, decoding schemes using dye bar codes are employed to register individual sensor placement.

Array Decoding

Individual sensor positions are different from array to array, so decoding methods are necessary to determine the position of each microsensor. Universal decoding schemes are based on spectrally encoding the microspheres with various fluorescent dyes (4-6). Each type of microsphere is tagged with multiple fluorescent dyes to form unique bar codes that differentiate the microsensors from one another. Microspheres can be labeled with dyes using either internal or external attachment. Internal dye entrapment is performed by swelling microspheres in the presence of dyes dissolved in an organic solvent; the dyes become trapped when the solvent is evaporated (5). External dye encoding is done via covalent attachment to the microsphere surface (5). Multiple encoding dyes are used, requiring only that their emission spectra differ from that of the target label. Dyes can also be incorporated at different distinguishable concentrations such that hundreds of different sensor types are possible.

Another decoding strategy for the DNA arrays exists where sequence determination is performed by multiple DNA hybridizations (33). This hybridization decoding scheme identifies each microsensor by its DNA probe sequence attached to the microsphere.

Oligonucleotide Probe Designs

One method to detect DNA hybridization is by using fluorescently-labeled DNA targets. The DNA probe sequences in this case are synthesized either prior to coupling to the microspheres or built base-by-base directly on the bead surface using standard phosphoramidite chemistry. The polymerase chain reaction (PCR) can amplify the desired target from the sample of interest, using polymerase enzymes that synthesize the sequence onto fluorescently-labeled primers. The primers are specific for the targets of interest and ensure that extraneous sequences are not amplified.

An alternative method that allows for screening unlabeled targets use DNA probes known as molecular beacons (4). The molecular beacon scheme contains multiple distinct functional components. A 5' biotin is incorporated for microsphere attachment and an oligonucleotide-hairpin structure serves as both the probe element and the fluorescent-signal generator. The center of the hairpin structure is designed to act as the specific probe, complementary to the oligonucleotide target of interest, as illustrated in Figure 6. Both ends of the molecular beacon's hairpin DNA sequence are complementary so as to form a loop configuration in the absence of target. A fluorophore is attached to one end of the DNA sequence and a quencher is attached to the other end (for example, fluorescein and 4-(4-dimethylaminophenylazo) benzoic acid respectively).

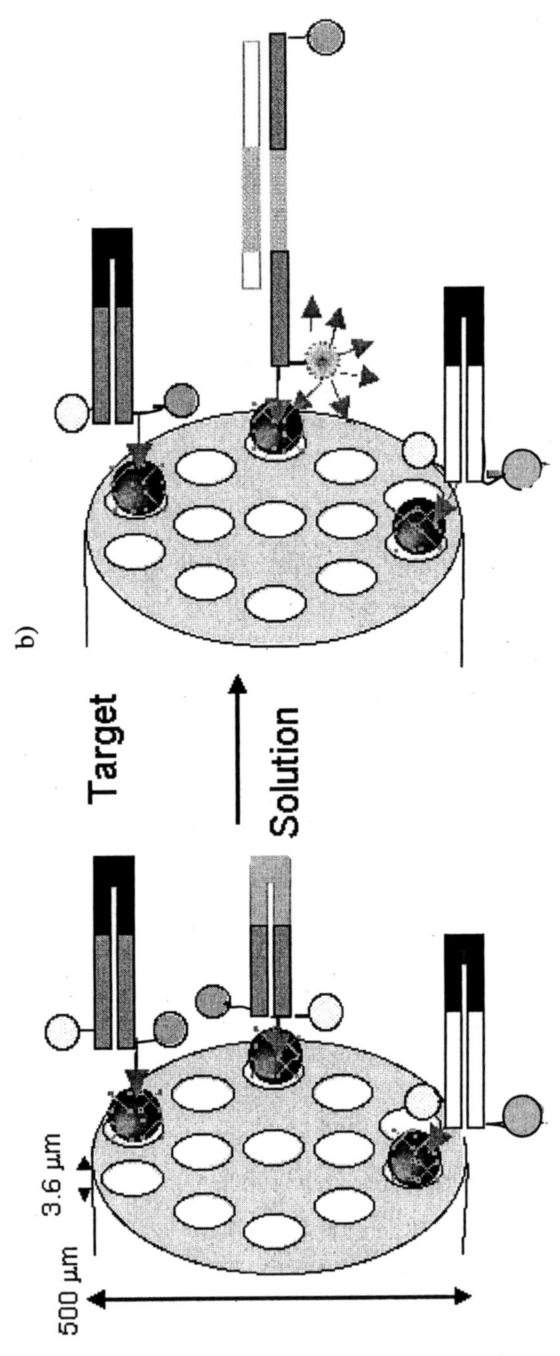
500 μm
3.6 μm
a)
b)
Target
Solution

c)

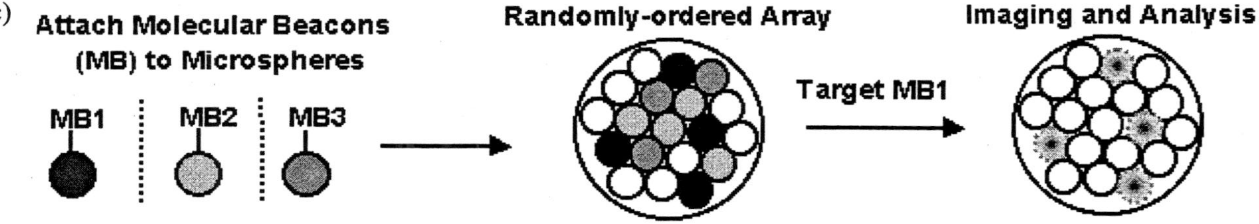

Figure 6. a) Diagram of a fiber-optic microarray with molecular beacon (MB) functionalized-microspheres randomly distributed on the fiber where the fluor and quencher are in proximity. b) The same fiber with the MB hybridized to its complement, causing the quencher to move away from the fluor and generating a signal. c) Diagram of the array containing three different MB probes and the result shown on the imaging system when complementary targets hybridize. The sensing regions are not drawn to scale. Tens of thousands of sensors can occupy an array.

When the hairpin configuration of the probe is formed, the fluor and quencher are brought together. In this conformation, the quencher reduces the signal generated by the fluor. In the presence of the correct nucleic acid target, the beacon unfolds and the fluor and quencher are separated so that the fluorescence is no longer inhibited.

In either method, complementary target binding increases fluorescence from the hybridized microsensor, which is transmitted back through the fiber to an ICCD camera. The difference between the background and the hybridization intensity levels correspond to the amount of target binding to the probe DNA. On the imaging system, the fiber position was fixed and all solutions were brought to the fiber's distal face. Due to the miniature size of the fiber (200-500 μm), sample volumes as low as 3 μL can be used, although 10 μL volumes were commonly used to avoid problems associated with solvent evaporation (34). Both sensor-types can be regenerated by rinsing with 90% formamide solution and reused over 50 times with minimal signal degradation (4).

For the fluorescent target method, overall hybridization times are relatively short, with 10 μL volumes of nanomolar target concentrations resolved in minutes and femtomolar (10^{-15} M) target concentrations able to be detected in less than an hour (29). This method has a limit of detection of 10^{-21} mol (ca. 600 molecules). For the molecular beacon scheme, a 14-fold increase in the mean intensity between the background and target-exposed array was seen in one hour for micromolar concentrations. The detection limit for the molecular beacon method is 10^{-15} mol (10^7 molecules).

The array can be custom designed for specific targets of interest. As new sequences are discovered, probes can be made readily to adapt the sensor , without affecting the array's sensing capabilities. Fiber optic DNA arrays are a viable alternative array platform with the ability to detect single base-pair variations and extremely low detection limits. In addition to these applications, the high-density microarray platform has the ability to use living cells as fluorescent biosensors (35).

Conclusions

Within the biological and chemical sensor community, a longstanding goal has been to make miniature arrays of different sensor elements that can simultaneously detect multiple analytes. Fiber-optic imaging bundles are an ideal platform for such sensor arrays because high-density arrays containing 6000-50,000 individual sensors can be made that have an overall diameter of just 1 mm. Each sensor is optically wired in the fiber bundle, allowing the sensors to be individually interrogated for continual or repetitive measurements. The characteristic redundancy of this sensor array allows for signal averaging across like sensors, increasing the signal-to-noise ratio and sensitivity of the array. The total internal reflection properties of optical fibers make them suitable for remote sensing applications such as explosives detection. The fiber-optic sensor array platform has already been applied to organic vapor and DNA detection problems. This technology presently is being applied to other solution-based analyses such as enzyme and protein assays.

Acknowledgement

This work was supported by The National Institutes of Health (GM48142), D.A.R.P.A., the Office of Naval Research, and The Department of Energy.

Literature Cited

(1) Wolfbeis, O. S. *Anal. Chem.* **2000**, 72, 81R-89R.
(2) These fiber optic microarray sensor technology is the basis for a commercial venture, Illumina (www.illumina.com).
(3) Pantano, P.; Walt, D. R. *Chem. Mater.* **1996**, 8, 2832-2835.
(4) Steemers, F. J.; Ferguson, J. A.; Walt, D. R. *Nat. Biotechnol.* **2000**, 18, 91-94.
(5) Michael, K. L.; Taylor, L. T.; Schultz, S. L.; Walt, D. R. *Anal. Chem.* **1998**, 70, 1242-1248.
(6) Dickinson, T. A.; Michael, K. M.; Kauer, J. S.; Walt, D. R. *Anal. Chem.* **1999**, 71, 2192-2198.
(7) Buck, L. B. *Cell* **2000**, 100, 611-618.
(8) Malnic, B.; Hirono, J.; Sato, T.; Buck, L. B. *Cell* **1999**, 96, 713-725.
(9) Sullivan, S. L.; Ressler, K. J.; Buck, L. B. *Curr. Opin. Genet. Dev.* **1995**, 5, 516-523.
(10) Dickinson, T. A.; White, J.; Kauer, J. S.; Walt, D. R. *Trends Biotechnol.* **1998**, 16, 250-258.
(11) Gardner, J.; Bartlett, P. Electronic Noses Principles and Applications; Oxford University Press: Oxford, 1999.
(12) Park, J.; Groves, W. A.; Zellers, E. T. *Anal. Chem.* **1999**, 71, 3877-3886.
(13) Grate, J. W.; Wise, B. M.; Abraham, M. H. *Anal. Chem.* **1999**, 71, 4544-4553.
(14) Shurmer, H. V.; Corcoran, P.; Gardner, J. W. *Sens. Actuators B* **1991**, 4, 29-33.
(15) Freund, M. S.; Lewis, N. S. *Proc. Natl. Acad. Sci. U.S.A.* **1995**, 92, 2652-2656.

(16) Persaud, K.; Dodd, G. *Nature* **1982**, 299, 352-355.

(17) Gardner, J. W.; Shurmer, H. V.; Tan, T. T. *Sens. Actuators B* **1992**, B6, 71-5.

(18) Gardner, J. W.; Shurmer, H. V.; Cocoran, P. *Sens. Actuators B* **1991**, 4, 117-121.

(19) Doleman, B. J.; Lonergan, M. C.; Severin, E. J.; Vaid, T. P.; Lewis, N. S. *Anal. Chem.* **1998**, 70, 4177-4190.

(20) Lonergan, M. C.; Severin, E. J.; Doleman, B. J.; Beaber, S. A.; Grubbs, R. H.; Lewis, N. S. *Chem. Mater.* **1996**, 8, 2298-2312.

(21) Severin, E. J.; Doleman, B. J.; Lewis, N. S. *Anal. Chem.* **2000**, 72, 658-668.

(22) Reichardt, C. *Chem. Rev.* **1994**, 94, 2319-2358.

(23) Buck, L. B. *Annu. Rev. Neurosci.* **1996**, 19, 517-544.

(24) White, J.; Dickinson, T. A.; Kauer, J. S.; Walt, D. R. *Anal. Chem.* **1996**, 68, 2191-2202.

(25) Dickinson, T. A.; White, J.; Kauer, J. S.; Walt, D. R. *Nature* **1996**, 382, 697-700.

(26) Albert, K. J.; Walt, D. R. *Anal. Chem.* **2000**, 72, 1947-1955.

(27) Johnson, S. R.; Sutter, J. M.; Engelhardt, H. L.; Jurs, P. C. *Anal. Chem.* **1997**, 69, 4641-4648.

(28) Walt, D. A. *Science* **2000**, 287, 451-452.

(29) Epstein, J. R.; Lee, M.; Walt, D. R. , unpublished.

(30) Healey, B. G.; Matson, R. S.; Walt, D. R. *Anal. Biochem.* **1997**, 251, 270-279.

(31) Ferguson, J. A.; Boles, T. C.; Adams, C. P.; Walt, D. R. *Nat. Biotech.* **1996**, 14, 1681-1684.

(32) Ferguson, J. A., Steemers, F.J., Walt, D.A. *Anal. Chem.* **2000**, 72, 5618-5624.

(33) Ferguson, J. A.; Walt, D. A. , unpublished.

(34) Bronk, K. S.; Michael, K. L.; Pantano, P.; Walt, D. R. *Anal. Chem.* **1998**, 70, 2750-2757.

(35) Taylor, L. C.; Walt, D. R. *Anal. Biochem.* **2000**, 278, 132-142.

INDEXES

Author Index

Subject Index